Applied Solid State Physics

APPLIED
SOLID STATE
PHYSICS

edited by
W. Low and M. Schieber

Weizmann Science Press
of Israel, Jerusalem

Plenum Press, New York

The papers presented in this volume are edited
versions of lectures given at the Bathsheva de Rothschild
Seminar on Applied Physics held at the Hebrew University of
Jerusalem, Israel, April 21 — May 14, 1968

Standard Book Number 306-30485-6
Library of Congress Catalog Card Number: 70-119901

The Weizmann Science Press of Israel, P. O. B. 801, Jerusalem
Plenum Press, A division of
Plenum Publishing Corporation,
227 West 17th Street,
New York, N. Y. 10011

ISBN 978-1-4684-1856-9   ISBN 978-1-4684-1854-5 (eBook)
DOI 10.1007/978-1-4684-1854-5

Color plate reprint at the Central Press, Jerusalem
Jacket design: Ruth Wiener

IV

CONTENTS

CONTRIBUTORS

B. CHALMERS
Division of Engineering and Applied Physics,
Harvard University, Cambridge, Massachusetts

W. D. DOYLE
Sperry Rand Corporation, Univac Division, Philadelphia, Pennsylvania

J. A. GIORDMAINE
Bell Telephone Laboratories, Inc., Murray Hill, New Jersey

ALBERT A. FRIESEM
Radar and Optics Laboratory, Institute of Science and Technology, The University of Michigan

R. A. LAUDISE
Bell Telephone Laboratories, Inc., Murray Hill, New Jersey

B. T. MATTHIAS
University of California, La Jolla, California, and Bell Telephone Laboratories, Murray Hill, New Jersey

S. P. S. PORTO
Departments of Physics and Electrical Engineering, University of Southern California, Los Angeles

MORTON B. PRINCE
Electro Optical Systems, Inc., Pasadena, California, and UNESCO Specialist in Scientific Instrumentation

A. ROSE
RCA Laboratories, Princeton, N. J.

PREFACE

This book is a collection of a set of lectures sponsored by the Bathsheva
de Rothschild Seminars. It deals with different aspects of applied physics
which are an outgrowth of fundamental research. The courses were given
by experts engaged in their respective fields.

These review articles are intended to fill a gap between the many
research papers that are appearing today in pure science on one hand,
and in applied science on the other hand. It is a bridge between these two
It aims at the specialist in applied physics, chemistry and engineering,
working in these specialized fields, as well as at the graduate student,
interested in solid solid state physics, chemistry and electrical engineering.

While this book contains a range of different topics, there is an underlying logic in the choice of the subject material. The first three articles,
by Drs. Giordmaine, Friesem and Porto, deal with modern applied optics,
which arise to a large extent from the availability of coherent and powerful
laser sources. Two articles deal with materials, in particular that of
Dr. Chalmers on the theory and principle of solidification and that of
Dr. Laudise on the techniques of crystal growth. The last three articles,
by Drs. Matthias, Doyle and Prince, are concerned with the use of
materials in fields of superconductivity, computer storage and semiconductor photovoltaic effects. Dr. Rose gives a definitive review on
human and electronic vision, an out-growth of life-long activity in this
field.

We are grateful to the Bathsheva de Rothschild Foundation and to the
Hebrew University of Jerusalem for the assistance in making it possible
to organize the School in Applied Physics, held in Jerusalem in 1968.

ירושלים תובב"א, תש"ל

VII

1/Parametric Optics

J. A. GIORDMAINE

1. INTRODUCTION

The purpose of this paper is to discuss three examples of parametric field interactions which have recently become important in optics: optical parametric oscillation, parametric fluorescence, and the stimulated Raman effect. By a parametric interaction we refer to a process in which the application of a field E_p leads to regenerative amplification of two fields E_s and E_i with a phase relation of the form

$$\psi_p - (\psi_s + \psi_i) = \pi/2 \qquad (1)$$

maintained among the fields. The quantities $E_{p,s,i}$ may in principle represent any classical fields. In quantum mechanics the elementary parametric process is an interaction of three bosons, e.g., the annihilation of a photon and the creation of a photon and an optical or acoustic phonon as in the stimulated Raman or Brillouin effect, the creation of a photon and a polariton as in the Raman effect involving an infra red active lattice vibration, or simply the creation of two photons as in optical parametric oscillation.

From a classical point of view[1] the mechanism of the parametric process and the inherent frequency and phase insensitivity of parametric gain are made clear by the following example. Consider the local interaction of an intense monochromatic plane light wave $E_p(\omega_p, \psi_p)$, the pump, with a second plane wave $E_s(\omega_s, \psi_s)$, the signal, having lower but otherwise arbitrary frequency ω_s, and arbitrary phase ψ_s The fields are allowed to interact through the nonlinear polarization of the medium, in particular a polarization component P quadratic in the fields,

$$P = \chi E^2 \qquad (2)$$

where χ is a tensor nonlinear optical coefficient. As a result of Eq. (2) the pump and signal fields generate polarization at the difference or idler frequency

$$\omega_i \equiv \omega_p - \omega_s. \qquad (3)$$

This paper was also presented at the International Summer School of Physics, Enrico Fermi, Varenna, Course 42. It appears in essentially the same form in the Proceedings of the International Summer School of Physics, Course 42, Quantum Optics, edited by R. J. Glauber, Academic Press, New York, 1969. Permission of Academic Press, Inc. to republish this paper is gratefully acknowledged.

As shown in Figure 1, the local polarization current density component at ω_i, $J(\omega_i) = \partial P(\omega_i)/\partial t$, can amplify an electric field $E_i(\omega_i)$ having the particular phase $\psi_i = \psi_p - \psi_s - \pi/2$. The field $E_i(\omega_i)$ together with $E_p(\omega_p)$ generates polarization at the signal frequency having the correct phase to amplify the applied signal field $E_s(\omega_s)$. This cycle is regenerative and leads to exponential gain for the fields E_s and E_i as shown in Section II. In order that the phase relationship Eq. (1) for gain be time independent, we require that

$$(\partial/\partial t)(\psi_p - \psi_i) = \omega_p - \omega_s - \omega_i = 0. \tag{4}$$

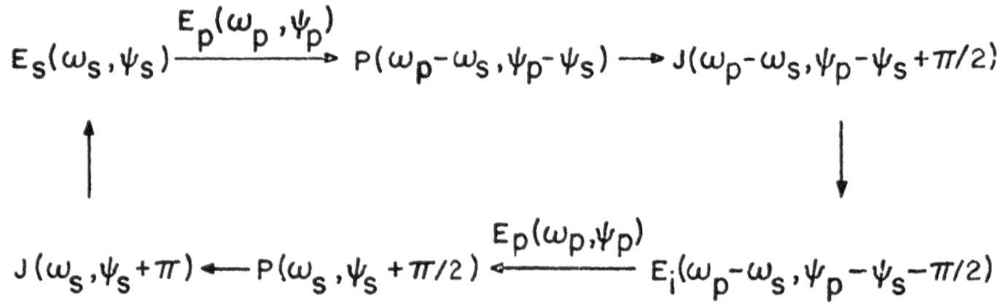

$$E_s(\omega_s,\psi_s) \xrightarrow{E_p(\omega_p,\psi_p)} P(\omega_p-\omega_s,\psi_p-\psi_s) \longrightarrow J(\omega_p-\omega_s,\psi_p-\psi_s+\pi/2)$$

$$\uparrow \qquad\qquad\qquad\qquad\qquad\qquad\qquad\qquad\qquad\qquad \downarrow$$

$$J(\omega_s,\psi_s+\pi) \longleftarrow P(\omega_s,\psi_s+\pi/2) \xleftarrow{E_p(\omega_p,\psi_p)} E_i(\omega_p-\omega_s,\psi_p-\psi_s-\pi/2)$$

FIGURE 1. Mechanism of local parametric amplification of signal and idler fields E_s and E_i in a medium having a polarization component P quadratic in the fields. The pump field E_p has given frequency ω_p and phase ψ_p. The signal and idler frequencies ω_s and ω_i satisfy $\omega_s + \omega_i = \omega_p$; the signal and idler phase ψ_s and ψ_i satisfy $\psi_p - (\psi_s + \psi_i) = \pi/2$. The polarization current density $J \equiv \partial P/\partial t$. The figure shows that a signal field of arbitrary frequency ω_s ($<\omega_p$) and phase ψ_s gives rise via the pump and idler fields to a current density $J(\omega_s,\psi_s + \pi)$ of just the proper phase to be amplified regeneratively. The arrows indicate the steps of the mixing and amplification processes.

From a quantum point of view[2-4] it is useful to express the interaction of the fields as a term in the field energy cubic in the fields. The nonlinearity appears in the Hamiltonian function for the fields as terms such as $\chi\, a_s^\dagger\, a_i^\dagger\, a_p$ describing annihilation of a pump photon and creation of a signal and an idler photon. This process has nonvanishing probability when Eq. (4), which expresses conservation of photon energy, is approximately satisfied.

The possibility of amplification of zero point field fluctuations leads to the observation of output from a parametric amplifier at the signal and idler frequencies even in the absence of an input. This emission, called parametric fluorescence or parametric noise, is the analog of fluorescence in a laser amplifier, and is discussed in Section III.

The parametric process differs from laser amplification in that: 1) the main features of the amplification process are describable classically, at least when the fields themselves are in coherent states and have classical properties[5], and 2) the gain is frequency insensitive, since the magnitude of χ in Eq. (2) is a weak function of frequency when the medium is transparent to ω_s and ω_i. This broadband gain makes possible optical amplifiers tunable over a large fraction of the visible spectrum.

2

In practice, however, the output frequency of an optical parametric oscillator is highly monochromatic, resembling the emission of a solid state laser. The frequency selectivity and tuning characteristics arise because macroscopic or nonlocal amplification requires significant gain over distances large compared with λ. Consider the interaction of plane wave pump, signal, and idler fields with wave vectors $\underset{\sim}{k}_p$, $\underset{\sim}{k}_s$, and $\underset{\sim}{k}_i$ respectively. In order that the phase relationship for gain Eq. (1) be valid everywhere, the condition

$$\underset{\sim}{\nabla} \cdot (\psi_p - \psi_s - \psi_i) = \underset{\sim}{k}_p - \underset{\sim}{k}_s - \underset{\sim}{k}_i = 0 \tag{5}$$

must be satisfied. From the quantum viewpoint Eq. (5) represents conservation of photon momentum. The condition for gain over a crystal length ℓ, $\mid \underset{\sim}{k}_p - \underset{\sim}{k}_s - \underset{\sim}{k}_i \mid < \pi/\ell$, is typically satisfied for a band of frequencies only a few cm^{-1} wide for a given value of $\underset{\sim}{k}_p$. Tuning occurs as the band of frequencies approximately satisfying Eq. (5) is shifted by temperature, angle, or electric field variations of the refractive index.

The principle of parametric amplification has long been familiar in mechanical and electrical oscillators for the phase sensitive subharmonic case $\omega_s = \omega_i = \omega_p/2$[6]. The extension of this process to nondegenerate tunable oscillation appears first to have been recognized ten years ago in the case of the ferromagnetic microwave amplifier[7].

Section II describes the theory of parametric gain for optical oscillators in their present form, and reviews currently available nonlinear optical materials. Section III discusses recent observations of parametric fluorescence. In Section IV the stimulated Raman effect is described as a parametric phenomenon, and the limiting short pulse behavior of a parametric amplifier is compared with the short pulse behavior of the laser.

II. OPTICAL PARAMETRIC AMPLIFICATION AND OSCILLATION

1. *Nonlinear Optical Crystals*

Crystals of greatest interest in parametric optical experiments have a) a large value of nonlinear optical coefficient χ [Eq. (2)], b) adequate birefringence to allow phase matching [Eq. (5)], c) low susceptibility to damage in intense pump light, and in addition such suitable physical properties as low optical absorption, etc. We now discuss each of these properties in turn.

(a) N o n l i n e a r O p t i c a l C o e f f i c i e n t s. We consider here materials having a macroscopic polarization component P quadratic in the optical electric field as in Eq. [8]. The formal theory of the nonlinear optical coefficients has been discussed in references [9] and [10]: Here we shall outline some quantitative aspects relevant to optical parametric phenomena.

For monochromatic optical fields the nonlinear coefficients are conventionally defined as in Eqs. (6) and (7), applicable to optical mixing and optical harmonic generation respectively.

$$P_i(\omega_3) = \sum_{jk} \chi_{ijk}(-\omega_3,\omega_1,\omega_2)E_j(\omega_1)E_k(\omega_2) \tag{6a}$$

$$P_i(\omega_2) = \sum_{jk} \chi_{ijk}(-\omega_2,\omega_3,-\omega_1)E_j(\omega_3)E_k^*(\omega_1) \tag{6b}$$

$$P_i(2\omega) = \sum_{jk} d_{ijk}(-2\omega,\omega,\omega)E_j(\omega)E_k(\omega) \tag{7}$$

The subscripts ijk indicate vector components along orthogonal crystal axes, and $\omega_3 = \omega_1 + \omega_2$. It follows from Eqs. (6) and (7) that

$$d_{ijk}(-2\omega,\omega,\omega) = \tfrac{1}{2}\chi_{ijk}(-2\omega,\omega,\omega).$$

The quantities χ_{ijk} can be shown to be real in the absence of absorption [11, 12]. The fields $P_i(\omega)$ and $E_i(\omega)$ in Eqs. (6) and (7) are amplitudes defined [13] by

$$P_i(t) = \tfrac{1}{2}[P_i(\omega)e^{-i\omega t} + c.c.].$$

Materials having nonvanishing χ_{ijk} occur in noncentrosymmetric crystal classes. Crystal symmetry leads to zero values for certain χ_{ijk} components and relations between others [14]. In addition permutation symmetry relations [11, 12]

$$\chi_{ijk}(-\omega_3,\omega_2,\omega_1) = \chi_{jik}(\omega_2,-\omega_3,\omega_1) = \chi_{kij}(\omega_1,-\omega_3,\omega_2) \tag{8}$$

imposed by the existence of an energy contribution cubic in the applied fields connect certain components which are unrelated by crystal symmetry. Where dispersive effects can be ignored, the symmetry relations of Eq. (8) reduce to $\chi_{ijk} = \chi_{jik} = \chi_{kij}$ [15].

Table I lists measured values of χ for a number of crystals of interest for parametric amplifiers and oscillators. Since d_{ijk} is symmetric in j and k it is usual to write $d_{ijk} = d_{i\ell}$ according to the rule $\ell = 1, 2, 3, 4, 5, 6$ for jk = 11, 22, 33, 23, 13, 12. For example in $LiNbO_3$

$$\begin{pmatrix} P_x \\ P_y \\ P_z \end{pmatrix} = \begin{pmatrix} 0 & 0 & 0 & 0 & d_{15} & 0 \\ -d_{22} & d_{22} & 0 & d_{15} & 0 & 0 \\ d_{31} & d_{31} & d_{33} & 0 & 0 & 0 \end{pmatrix} \begin{pmatrix} E_x^2 \\ E_y^2 \\ E_z^2 \\ 2E_yE_z \\ 2E_xE_z \\ 2E_xE_y \end{pmatrix} \tag{9}$$

4

TABLE I. Properties of Certain Nonlinear Optical Crystals

Crystal	Point Group	Nonlinear Optical Coefficient (Units of 10^{-9} esu) $d_{ijk}(-2\omega,\omega,\omega)$	ijk	Wavelength of Measurement $(2\pi c/\omega)$ λ	Ref.
KH_2PO_4	$\bar{4}$2m	1.4	312	1.06	(a)
		1.4	123		
$LiNbO_3$	3m	9	222	1.06	(b)
		16	311		
		1.1×10^2	333		
$K_{.6}Li_{.4}NbO_3$	4mm	21	311	1.06	[31]
		21	113		
		39	333		
HIO_3	222	24	123	1.06	(c)
$Ba_2NaNb_5O_{15}$	mm2	40	311	1.06	[32]
		46	322		
		47	333		
Ag_3AsS_3	3m	41	311	1.15	(d)
		68	222		
Te	32	1.3×10^4	111	10.6	(e)

(a) G.E.FRANCOIS, Phys. Rev., 143, 597 (1966); see [24] for comparison with quartz.
(b) G.D.BOYD, R.C.MILLER, K.NASSAU, W.L.BOND, and A.SAVAGE, Appl. Phys. Letters, 5, 234, (1964); R.C.MILLER and A.SAVAGE, Appl. Phys. Letters, 9, 169 (1966)
(c) S.K.KURTZ, and J.G.BERGMAN, Appl. Phys. Letters, 12, 186 (1968); Bull. Am. Phys. Soc., 13, 388 (1968)
(d) K.F.HULME, P.H.DAVIES, M.V.HOBDEN, Appl. Phys. Letters, 10, 133 (1967)
(e) C.K.N.PATEL, Phys. Rev. Letters, 15, 1027 (1965)

The theory of the nonlinear optical coefficients has been widely discussed [9] and formal expressions for χ are available in terms of dipole matrix elements [10, 11, 16-20]. In general however, the rigorous expressions for χ have not yet proved amenable to detailed calculation in particular cases. Useful understanding has come from classical or semiclassical anharmonic oscillator models of the nonlinear materials [21-23].

An important empirical relation, Miller's rule [24] correlates the nonlinear optical coefficients with the linear optical properties. This relation is

$$d_{ijk}(-2\omega,\omega,\omega) = \Delta\chi_{ii}(2\omega)\chi_{jj}(\omega)\chi_{kk}(\omega) \qquad (10)$$

where $\chi_{ii}(\omega)$ is the linear susceptibility equal to $\left(n_i^2 - 1\right)/4\pi$ at frequency ω, n_i is the refractive index for light polarized along the i axis, and Δ is approximately 2×10^{-6} esu. Almost all known nonlinear optical crystals have values of Δ within a factor of 4 of 2×10^{-6} esu, although the coefficients themselves vary over a range of $\sim 10^4$. Eq. (10) has been found valid in both the visible and infrared.

Eq. (10) can be understood in terms of an anharmonic oscillator model [21]. For example if the anharmonic term in the classical potential is assigned to have a magnitude equal to the harmonic term when the electronic displacement becomes equal to the lattice spacing, the model leads directly to Eq. (10) with the correct order of magnitude for Δ. Eq. (10) and Δ are also predicted by a semiclassical single electron model in which the optical properties are represented by a two-level system, and the nonlinearity estimated in terms of the birefringence.

Miller's rule has been a highly useful guide in the search for new nonlinear optical materials since it allows preliminary selection solely on the basis of the measured refractive indices.

(b) Optical Birefringence. Satisfaction of the frequency and k-vector matching conditions Eqs. (4) and (5) is not possible in isotropic crystals having normal dispersion [25]. In these materials $n_p > n_s, n_i$, and for a pump wavelength ~ 5300 Å and $\omega_s \approx \omega_i$ the mismatch $\Delta k = k_p - k_s - k_i = \frac{1}{c}(n_p \omega_p - n_s \omega_s - n_i \omega_i)$ is of the order of 10^4 cm^{-1} Parametric gain in this case is limited to an effective length of $\sim 2\pi/\Delta k \sim 10^{-3}$ cm.

However in an anisotropic crystal with adequate birefringence the pump wave can be given the polarization direction of lower refractive index, and one or both the signal and idler waves the polarization of higher refractive index, in this way compensating for normal dispersion[26, 27]. Since the refractive index is a function of propagation direction the desired oscillation frequency can be tuned by suitably orienting the crystal relative to the resonator to allow the desired frequencies to satisfy Eqs. (4) and (5).

The particular propagation directions normal to the optic axis in uniaxial crystals, or along a principal axis in biaxial crystals, are especially desirable[28]. In these directions, since double refraction is absent and the Poynting vectors of pump, signal, and idler waves having collinear k vectors are themselves collinear, a maximum interaction length is possible[13]. In addition the absence of double refraction allows the efficient use of highly focused pump light making possible oscillation at a much reduced threshold.

(c) Optical Damage. Some crystals useful in studies of optical parametric amplification show optically induced refractive index inhomogeneities in focused laser beams with power in the milliwatt range [29]. This effect occurs strongly in LiNbO$_3$ where the optical damage is severe enough to make continuously pumped optical oscillators impossible at room temperature. The damage in LiNbO$_3$ is at least partly an integrating effect and does not prevent pulsed laser pump experiments. The damage is erased and does not occur at temperatures above 170°C; it is produced by visible but not infrared light. A proposed mechanism [30] for the damage is photoexcitation of carriers from deep traps. The carriers are believed to drift out of the illuminated region under the effect of internal fields. On reaching the edge of the beam the carriers again drop into traps, giving rise to a space charge field which produces the refractive index inhomogeneity through the electrooptic effect.

6

Materials having "filled structures" such as $Ba_2 NaNb_5 O_{15}$ tend not to exhibit optical damage[31, 32]. These materials have no unfilled metal ion sites and hence are less susceptible to deep trap formation and resulting optical damage.

2. *Optical Parametric Gain*

As a specific example of parametric gain consider the propagation of monochromatic plane wave pump light with wavelength 5300 Å in $LiNbO_3$ along the $\hat{x}$ axis. The pump light is polarized in the direction of the $\hat{z}$ (optic) axis. We calculate the interaction of the extraordinary pump wave (ω_3) with collinear ordinary signal and idler waves $(\omega_1$ and $\omega_2)$ polarized in the $\hat{y}$ direction[33, 34]. The pump, signal, and idler fields are respectively

$$\underline{E}_p = \frac{1}{2} \left[E_3(x)e^{i(k_3 x - \omega_3 t)} + c.c. \right] \hat{z}$$

$$\underline{E}_s = \frac{1}{2} \left[E_1(x)e^{i(k_1 x - \omega_1 t)} + c.c. \right] \hat{y} \tag{11}$$

$$\underline{E}_i = \frac{1}{2} \left[E_2(x)e^{i(k_2 x - \omega_2 t)} + c.c. \right] \hat{y}$$

It is assumed that

$$\omega_3 - \omega_1 - \omega_2 = 0 \tag{12}$$

but that in general

$$k_3 - k_1 - k_2 \equiv \Delta k \neq 0 \tag{13}$$

From Eq. (2) the nonlinear polarization components at ω_1 and ω_2 have the form [35]

$$P_s = \frac{1}{2} \left[P_1 e^{i(k_3 - k_2)x - \omega_1 t} + c.c. \right]$$

$$P_i = \frac{1}{2} \left[P_2 e^{i(k_3 - k_1)x - \omega_2 t} + c.c. \right] \tag{14}$$

where

$$P_1 = 2d_{15} E_3 E_2^*$$

$$P_2 = 2d_{15} E_3 E_1^* \tag{15}$$

Additional nonlinear polarization occurs at ω_3 proportional to $E_1 E_2$ and in fact provides the mechanism of pump depletion. This problem is discussed in detail in [36] and can be ignored in the present discussion of small signal gain. We substitute Eqs. (11), (14) and (15) into the wave equation (16)

7

$$\nabla \times (\nabla \times \underset{\sim}{E}) + \frac{1}{c^2} \frac{\partial^2 \underset{\sim}{D}}{\partial t^2} = - \frac{4\pi^2}{c^2} \frac{\partial^2 P}{\partial t^2}, \tag{16}$$

where linear polarization is included in D, and P represents only nonlinear polarization. Let n_1^0 and n_2^0 be the ordinary signal and idler refractive indices. In the usual approximation that $dE_i(x)/dx \ll E_i/\lambda$, it follows that

$$\frac{dE_1}{dx} = \frac{4\pi i \omega_1 d_{15}}{cn_1^0} e^{i\Delta kx} E_3 E_2^*$$

$$\tag{17}$$

$$\frac{dE_2^*}{dx} = \frac{-4\pi i \omega_2 d_{15}}{cn_2^0} e^{-i\Delta kx} E_3^* E_1$$

Solutions of Eqs. (17) have the form

$$E_1 = (E_{1+} e^{sx} + E_{1-} e^{-sx}) e^{i\Delta kx/2}$$

$$E_2^* = \left(E_{2+}^* e^{sx} + E_{2-}^* e^{-sx} \right) e^{-i\Delta kx/2} \tag{18}$$

where the amplitude gain s is given by

$$s = [\gamma^2 - (\Delta k/2)^2]^{\frac{1}{2}}. \tag{19}$$

Maximum gain $s = \gamma$ occurs when $\Delta k = 0$, and is given by [35]

$$\gamma = \left(\frac{16\pi^2 \omega_1 \omega_2 d_{15}^2 E_3 E_3^*}{c^2 n_1^0 n_2^0} \right)^{\frac{1}{2}} \tag{20}$$

The phase matched gain γ is related to the pump intensity I through the relation $I = n_3^e c E_3 E_3^* /8\pi$. For the case of LiNbO$_3$ with pump wavelength 5300 Å in air and nearly degenerate amplification, $\omega_1 \approx \omega_2$, one predicts on the basis of the value of $d_{15} = d_{31}$ given in Table I and the refractive indices $n_1^0 = n_2^0 = 2.23$ a gain

$$\gamma = 2.9 \times 10^{-4} [I(w \, cm^{-2})]^{\frac{1}{2}} \, cm^{-1}. \tag{20}$$

The available gain γ is proportional to the square root of the pump intensity and to the nonlinear coefficient d, and decreases slowly from its maximum value as ω_1 and ω_2 depart from degeneracy.

Some recent experimental observations of parametric gain are described in Table II and Figures 2 and 3.

8

TABLE II. Measurements of Parametric Gain

Crystal	Pump Wavelength μ	Signal Wavelength μ	Pump Power	Crystal Length cm	Gain dB	Ref.
ADP	.348	0.633	2×10^6 W	8	0.7	(g),(l)
LiNbO$_3$	.515	1.15	6×10^{-3} W	0.84	9×10^{-7}	(h)
KDP	.530	1.06	10^8 W cm^{-2}	3	0.4	(i)
ADP	.347	0.694	5×10^6 W cm^{-2}	3	9×10^{-3}	[55]
Te	10.6	17.9	10 W	0.1	11	(j)
LiNbO$_3$	.530	1.15	5×10^4 W	1	0.12	(k)

(g) C.C. WANG and G.W. RACETTE, App. Phys. Letters, 6, 169 (1965); Physics of Quantum Electronics, Ed. by P.L. Kelley et al., McGraw-Hill Book Co., New York, p. 20 (1966)

(h) [37]. Note that for small amplification the effective gain γ in a parametric oscillator experiment is the square root of the amplification measured in a mixing experiment, so that the quoted amplification is quite significant.

(i) S.A. AKHMANOV, A.I. KOVRIGIN, A.S. PISKARSKAS, V.V. FADEYEV, and R.V. KHOKHLOV, ZhETF Pis'ma, 2, 300 (1965); Soviet Phys. JETP Letters, 2, 191 (1965)

(j) C.K.N. PATEL, Appl. Phys. Letter, 9, 332 (1966)

(k) M.J. COLLES and R.C. SMITH, Appl. Phys. Letters, 10, 309 (1967)

(l) Other gain measurements by the indirect mixing method have been reported by N. VAN-TRAN, J. SPALTER, J. HANUS, J. ERNEST, and D. KEHL, Phys. Letters, 19, 285 (1965)

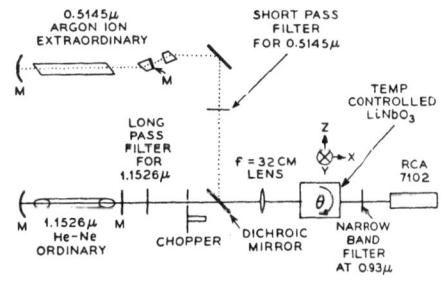

PARAMETRIC DIFFERENCE MIXING EXP.

FIGURE 2. Mixing of 5145 Å signal and CW 11526 Å pump light to generate difference frequency idler at 9299 Å in LiNbO$_3$ (after Boyd and Ashkin, [37]. The idler intensity measurement gives the gain at 5145 Å.

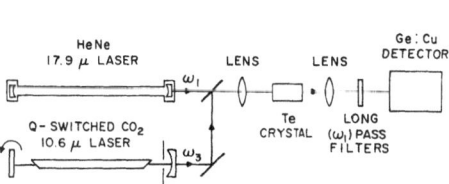

FIGURE 3. Tellurium parametric amplifier of 17.9μ infra red light with Q-switched 10.6μ CO$_2$ laser pump (after Patel, Ref. [j]).

3. Optical Parametric Oscillation

We consider initially the example of a LiNbO$_3$ optical resonator having plane end mirrors highly reflecting at ω_1 and ω_2; we represent the cavity modes in the one-dimensional standing plane wave approximation, with the propagation and polarization directions described in Part 3 (Figure 4).

The plane wave approximation is valid in calculating the gain in a short plane-parallel resonator at the focal point of a very lightly focussed pump beam [13]; however as will be seen the use of a plane-parallel resonator is highly inefficient. The extension of the threshold calculations described here to the general gaussian beam case is given in [37].

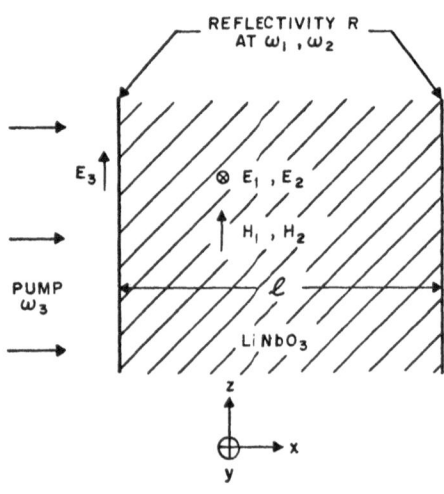

FIGURE 4. Plane wave resonator geometry for parametric oscillation in LiNbO₃.

From the gain considerations in Part 2 we expect that for purely travelling wave pump light the oscillation threshold will occur when the signal-idler gain per single pass across the resonator equals the resonator losses for a round trip around the resonator. Let R be the power reflection coefficient, assumed equal for the signal and idler, at each end reflector. Then the threshold condition is given by

$$e^{2\gamma\ell}\, R^2 = 1 \qquad (22)$$

For the usual case of small losses, $R \approx 1$, Eq. (22) becomes

$$\gamma \approx \frac{(1-R)}{\ell} \qquad (23)$$

For R = 0.99 and $\ell = 1$ cm, Eqs. (20) and (22) predict a threshold pump intensity $I = 1.2 \times 10^3$ wcm⁻². In fact, however, the distribution of longitudinal modes plays a crucial role in determining the oscillation frequency and threshold. As shown below, the threshold predicted by Eq. (22) requires the occurrence of two longitudinal modes resonant at ω_1 and ω_2.

Consider a traveling wave pump

$$\underset{\sim}{E}_p = E_3 \sin(k_3 x - \omega_3 t)\hat{z}. \qquad (24)$$

We wish to describe the coupling of $\underset{\sim}{E}_p$ with two particular resonator modes having electric and magnetic fields of the form

$$\hat{y}E_1(t)\sin k_{10}\, x, \qquad \hat{z}H_1(t)\cos k_{10}x$$

and $\hspace{9cm} (25)$

$$\hat{y}E_2(t)\sin k_{20}x, \qquad \hat{z}H_2(t)\cos k_{20}x.$$

To satisfy boundary conditions $E_1 = E_2 = 0$ at the mirror surfaces $x = 0, \ell$ we require that

$$k_{10} = \frac{m_1 \pi}{\ell}, \qquad k_{20} = \frac{m_2 \pi}{\ell}, \qquad (26)$$

10

where m_1 and m_2 are integers. The resonant frequencies of the two modes are

$$\omega_{10} = \frac{ck_{10}}{n_1^0}, \quad \omega_{20} = \frac{ck_{20}}{n_2^0} \tag{27}$$

In general, the resonant frequencies will not sum to ω_3,

$$\delta\omega \equiv \omega_3 - \omega_{20} - \omega_{10} \neq 0,$$

and

$$\delta k \equiv k_3 - k_{20} - k_{10} \neq 0.$$

It is useful to define the normal mode amplitudes, e.g.

$$a_1(t) = E_1(t) + iH_1(t)/n_1^0$$

and $$\tag{28}$$

$$a_1^*(t) = E_1(t) - iH_1(t)/n_1^0$$

From Eqs. (11) and (24)-(28) the nonlinear polarization component at ω_2 can be calculated, and its spatial Fourier sine component of period $2\pi/k_{20}$ is found to be

$$P_2(t) = \frac{d_{15}}{2} E_3 \sin k_{20}x \left[a_1 e^{i\omega_3 t} \frac{(e^{-i\delta k\ell} - 1)}{\delta k\ell} + c.c. \right] \tag{29}$$

All other Fourier components are orthogonal to $a_2(t)$. An expression similar to Eq. (29) gives $P_1(t)$ the polarization component which interacts with $a_1(t)$.

Substitution of Eqs. (28) and (29) into Maxwell's equations leads to the coupled mode equations (30) for $a_1(t)$ and $a_2(t)$.

$$\frac{da_1}{dt} = - i\omega_{10}a_1 + \frac{2\pi i d_{15}\omega_1 E_3}{(n_1^0)^2} \left(\frac{e^{i\delta k\ell} - 1}{\delta k\ell} \right) a_2^* e^{-i\omega_3 t} - K_1 a_1$$

$$\tag{30}$$

$$\frac{da_2^*}{dt} = i\omega_{20}a_2^* - \frac{2\pi i d_{15}\omega_2 E_3^*}{(n_2^0)^2} \left(\frac{e^{-i\delta k\ell} - 1}{\delta k\ell} \right) a_1 e^{i\omega_3 t} - K_2 a_2^*$$

The damping terms $K_i a_i$ have been added phenomenologically and are adequate to represent small losses at the mirrors.

$$K_i = \frac{c(1-R)}{2\ell n_i^0} \tag{31}$$

We shall make the approximation that the signal and idler losses are the same, $K_1 = K_2 \equiv K$, and that $n_1^0 = n_1^0 \equiv n^0$. For this case the solutions of eq. (30) have the simple form

$$a_1(t) = a_{10}e^{-i\omega_{10}'t+\sigma t}$$

$$\overset{*}{a_2}(t) = \overset{*}{a_{20}}e^{i\omega_{20}'t+\sigma t}$$

where

$$\sigma = -K \pm \frac{1}{2}\left[\frac{c^2\gamma^2}{(n^0)^2}\frac{\sin^2(\delta k\ell/2)}{(\delta k\ell/2)} - (\delta\omega)^2\right]^{\frac{1}{2}}, \tag{32}$$

γ is the spatial gain given by Eq. (20), and the oscillation frequencies ω'_{10} and ω'_{20} are

$$\omega_{10}' = \omega_{10}' + \delta\omega/2$$

$$\omega_{20}' = \omega_{20}' + \delta\omega/2 \tag{33}$$

satisfying $\omega'_{10} + \omega'_{20} = \omega_3$.

Threshold for oscillation occurs when $\sigma = 0$. For the optimum case of signal and idler resonance $\delta\omega = 0$, $\delta k = 0$, the threshold condition is found from Eqs. (30) and (32) to occur when the gain γ, given by Eq. (30), is equal to

$$\gamma_0 = (1-R)/\ell.$$

This result was predicted by Eq. (23) from gain considerations.

When $\delta\omega \neq 0$ and $\delta k \neq 0$ the required gain γ for threshold is given by Eq. (34),

$$\frac{\gamma^2}{\gamma_0^2} = \frac{1 + (\delta\omega/\delta\omega_c)^2}{\sin^2(\delta k\ell/2)/(\delta k\ell/2)^2} \tag{34}$$

which is the ratio of the threshold pump power at $(\delta\omega, \delta k)$ to its value at $\delta\omega = 0$, $\delta k = 0$. In Eq. (34) $\delta\omega_c = 2K$ is the full longitudinal mode width at half intensity.

The significance of Eq. (34) can be seen from Figure 5 which shows a representative distribution of signal and idler modes near the frequencies of maximum gain $\Delta k = 0$. Possible modes with frequencies ω'_{10} and ω'_{20} satisfying Eq. (33) lie within 1/4 the longitudinal mode spacing (typically 10^{-1} cm^{-1}) of the $\Delta k = 0$ frequencies. However Eq. (34) shows that the nearest possible modes may have a high threshold if $\delta\omega/\delta\omega_c \gg 1$. As a result oscillation may occur with minimum threshold in a pair of modes substantially separated from the $\Delta k = 0$ frequencies. In oscillators recently demonstrated the resulting frequency uncertainty has been as large as 10^1 to 10^2 cm^{-1} [33, 34].

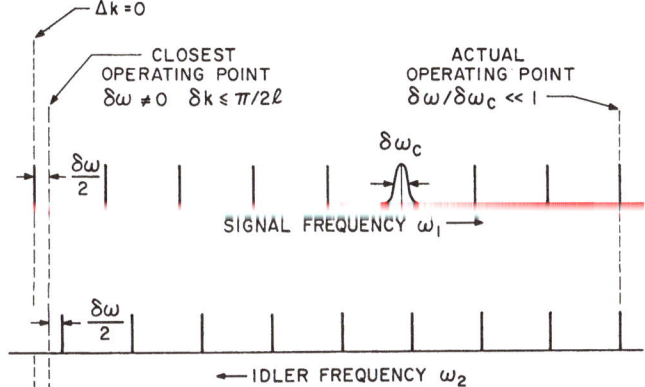

FIGURE 5. Schematic signal and idler longitudinal mode distribution; signal and idler frequencies satisfying $\omega_s + \omega_i = \omega_p$ fall on a vertical line in this diagram. Maximum gain occurs at the phase matched frequencies where $\Delta k = 0$; minimum threshold power (actual operating point) may occur at the nearest coincidence of signal and idler modes.

The characteristics and tuning range of several recent experimental optical oscillators are shown in Table III. The experiments to date (1968) have been performed with pulsed laser pump light.*

TABLE III. Characteristics of Experimental Pulsed Parametric Oscillators

Crystal	Pump Wavelength μ	Oscillator Tuning Range μ	Total Output Power W	Conversion Efficiency %	Tuning Mechanism	Ref.
LiNbO$_3$	0.529	0.97-1.15	30	0.4	Temp.	[33, 34]
KDP	0.530	0.96-1.18	3×10^3	0.04	Rotation	m
LiNbO$_3$	0.530	0.73-1.193	10^3	0.1	Temp.	n
LiNbO$_3$	0.530	0.68-2.36	50	0.1	Rotation	o
LiNbO$_3$	0.694	1.0-1.08	4×10^4	1	Elec. field, rotation	[41]

(m) S. A. AKHMANOV, A. I. KOVRIGIN, V. A. KOLOSOV, A. S. PISKARSKAS, V. V. FADEYEV, and R. V. KHOKHLOV, ZhETF Pis'ma, **3**, 372 (1966); Soviet Phys., **3**, 372 (1966)

(n) J. A. GIORDMAINE and R. C. MILLER, Appl. Phys. Letters, **9**, 298 (1966)

(o) R. C. MILLER and W. A. NORDLAND, Appl. Phys. Letters, **10**, 53 (1967)

* Note added in proof: CW optical parametric oscillation has recently been achieved by R. G. Smith et al., Appl. Phys. Letters, **12**, 308 (1968) and by R. L. Byer et al., ibid., **13**, 109 (1968).

With a pump wavelength of 5300 Å tuning has been demonstrated over a range of 6840 to 23550 Å, 70% of the theoretical range of the oscillator [38]. Figure 6 shows a typical experimental arrangement for a LiNbO$_3$ optical oscillator [33].

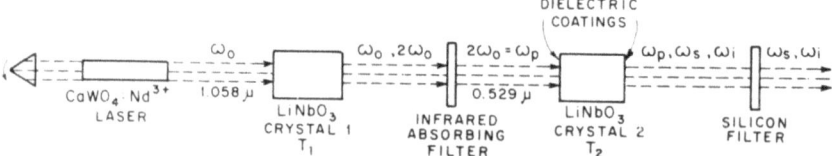

FIGURE 6A. Pulsed LiNbO$_3$ parametric oscillator, pumped at 5290 Å (after Giordmaine and Miller [33,34]). Crystal 2 is the oscillator, with dielectric coated reflectors; crystal 1 is a harmonic generator. Oscillation in LiNbO$_n$ has been observed in the range 6840 to 23550 Å.

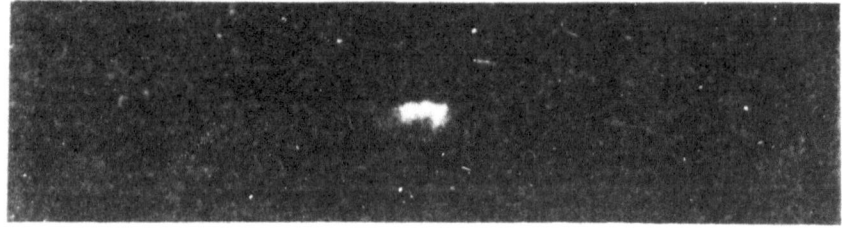

1 A

FIGURE 6B. Spectrum of LiNbO$_3$ parametric oscillator, near 9600 Å (J. A. Giordmaine and R. C. Miller, unpublished). The oscillator longitudinal mode spacing is 0.4 Å. The spectra show pulsed oscillation in a single longitudinal signal mode (above) and in three adjacent modes (below).

Threshold pump powers predicted on the basis of Eq. (23) are found to give order of magnitude agreement with observed thresholds [33-35]. Conversion efficiencies of $\gtrsim 1\%$ observed to date are less than expected on the basis of highly coherent pump light [36]; the discrepancy is undoubtedly connected with the complex multimode solid state laser sources which have so far been used in the experiments. *

4. Optimum Design of an Optical Parametric Oscillator

The plane wave analysis above predicts that a crystal of $LiNbO_3$ of length $\ell = 1$ cm with 1% end mirror losses (R = 0.99) should oscillate with a threshold pump intensity of 1.2×10^3 w cm^{-2} at 5300 Å. This analysis is valid for a single mode gaussian laser pump beam lightly focused at the crystal. The condition for a 1 cm crystal is that the beam diameter $2w_0 \geq \sqrt{8\lambda\ell}$ or 3×10^{-2} cm, and implies a minimum pump power of ~ 1 w for a plane mirror resonator.

Recent analyses [13, 37] show that a substantial reduction in threshold is obtained by the use of focused beams. For the case of $LiNbO_3$ with a single mode gaussian laser pump at 5147 Å, a crystal of length 1 cm and 1% end mirror losses is found to have a minimum threshold pump power of 2×10^{-2} w under optimum focusing conditions (Figure 7). Minimum pump power is obtained with mirrors of radius 0.56 cm on the surface of the crystal.

The low threshold power required for a well-designed parametric oscillator shows the feasibility of continuously pumped oscillators.

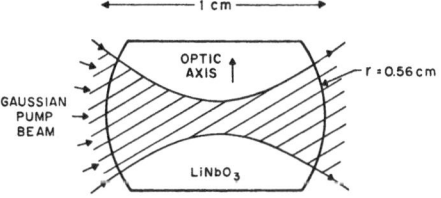

FIGURE 7. Optimum geometry for 1 cm $LiNbO_3$ oscillator having minimum threshold, as calculated by Boyd and Kleinman [13]. The reflectors have radius 0.56 cm and are coated on the crystal surface. The threshold 5147 Å pump power at optimum focusing for the case of 99% reflectance is 22 mW.

5. Tuning

Since parametric amplification occurs at those frequencies satisfying the phase matching conditions Eqs. (4) and (5), the oscillator can in

---

* Note added in proof: Efficiency as high as 38% have recently been reported by Kreuter, Appl. Phys. Letters, **13**, 57 (1968) and Bjorkholm, ibid., **13**, 53 (1968), approaching theoretical expectations.

principle be tuned by any change of conditions affecting the refractive index. Tuning has been demonstrated by temperature variation [33, 34, 39], crystal rotation [38, 40, 41], and application of electric field [41]. Figs. 8, 9, 10 show examples of tuning by these methods. Ferroelectric crystals such as $LiNbO_3$ display an especially large temperature coefficient of refractive index and thus a highly temperature dependent oscillator frequency. Near degeneracy a 1°C temperature variation leads to a signal wavelength change of ~300 Å in $LiNbO_3$ pumped at 5300 Å.

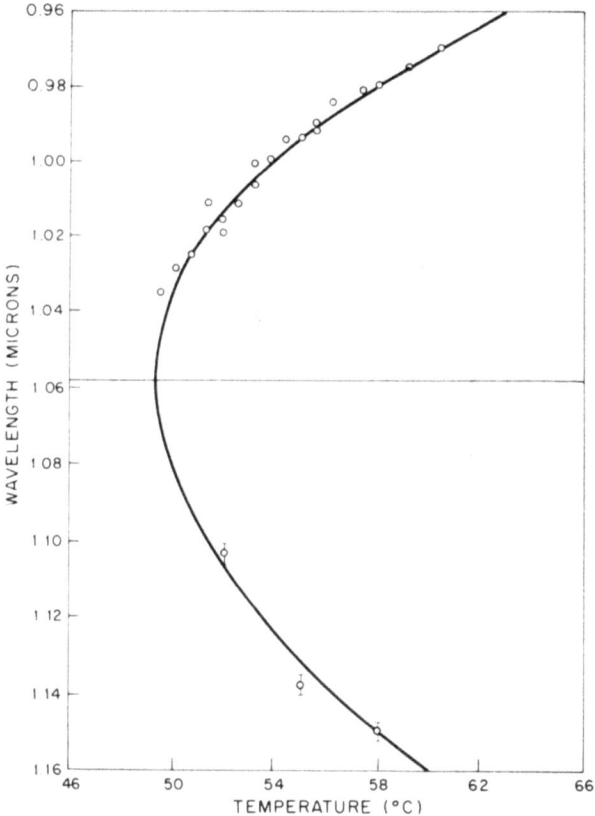

FIGURE 8. Temperature tuning of a $LiNbO_3$ parametric oscillator near degeneracy (1. 058 µ) (after Giordmaine and Miller [33, 34]). The pump wavelength is 5290 Å.

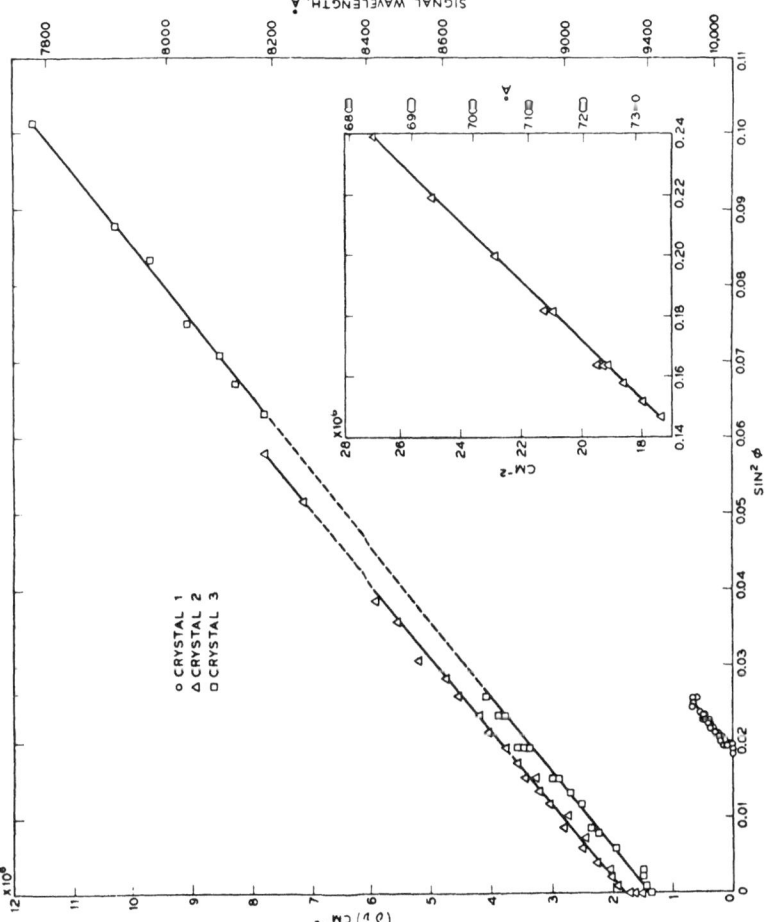

FIGURE 9. Crystal rotation tuning of signal frequency of a LiNbO₃ parametric oscillator (after Miller and Nordland, [38]). The pump wavelength is 5300 Å. δ⊽ designates the shift in cm⁻¹ from the degenerate frequency 9434 cm⁻¹. The quantity φ is the complement of the internal angle between the optic axis and the pump phase propagation direction.

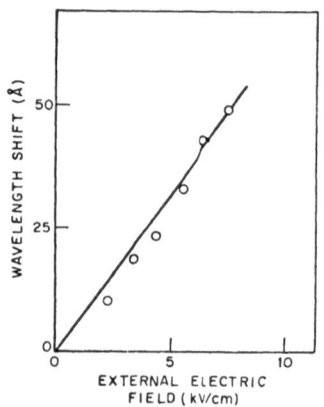

FIGURE 10. Electric field tuning of a LiNbO$_3$ parametric oscillator (after Kreuzer, [41]) near 10300 Å. The pump wavelength is 6943 Å, the tuning sensitivity 6.2 A kV^{-1}cm.

6. *Brief Historical Summary*

Following the discovery of harmonic generation in 1961 by P. A. Franken et al. [8] independent proposals for optical parametric amplifiers and oscillators were made by Kingston [42], Kroll [43], Akhmanov and Khokhlov [44], and Giordmaine and Kleinman [45]. The mechanism was discussed in detail by Armstrong, Bloembergen, Ducuing and Pershan [11]. A review of recent experimental work is given in Tables II and III.

Other discussions of the amplifier and oscillator theory have been given by Akhmanov and Khokhlov [46], by Akhmanov, Dmitriev, Modenov, and Fadeev [47], and by Grigor'ev, Rubenko, and Khokhlov [48]. The parametric coupling of optical modes has been treated by Kingston and McWhorter [49], and by Yariv [50]. Harris has disscussed the effects of multifrequency pumping on parametric oscillators [51], and has proposed interesting backward wave optical oscillators [52].

III. OPTICAL PARAMETRIC NOISE

The classical parametric amplification and oscillator theory as described in Section II is inadequate to discuss the weak light emission observed at the signal and idler frequencies with the pump below threshold, or in the complete absence of feedback. The emission, called optical parametric noise or optical parametric fluorescence, has recently been observed [53-58]. The same phenomenon contributes to noise in microwave parametric amplifiers, and the theory has been extensively discussed in the context of cavity modes [2, 59-61]. More recently scattering theories particularly appropriate to noise at optical frequencies have been developed [62-65]. References [64] and [65] contain a critical summary of optical parametric noise theory. The statistical properties of both the noise and oscillator output are discussed in [3] and [4].

In this lecture we derive one of the principal results of the noise theories in a nonrigorous heuristic way, making use of the fact [2, 59] that the output noise power is equivalent to the stimulated emission produced by one photon per mode of the radiation field.

Consider the case of parametric noise emitted close to the direction of plane wave pump light. We discuss a crystal of LiNbO$_3$ of length ℓ, aperture area A, with the pump light propagating along $\hat{x}$. Under phase matched conditions $\Delta k \ell \ll 1$, the classical Eqs. (18) for the amplified signal and idler fields $E_1(x)$ and E_2^* become

$$E_1(x) = E_1(0)\cosh sx + i\beta E_2^*(0)\sinh sx$$

$$E_2^*(x) = E_2^*(0)\cosh sx - i\beta^{-1}E_1(0)\sinh sx \tag{35}$$

where $\beta = (E_3 n_2\omega_1/E_3^*n_1\omega_2)^{\frac{1}{2}}$, E_3 is the pump field amplitude, s is given by Eqs. (19) and (20), and n_1 and n_2 are the refractive indices at ω_1 and ω_2 respectively.

At low gain, $sx \ll 1$, and in the absence of an input signal, $E_1(0) = 0$, it follows from Eq. (35) that the signal output at $x = \ell$ is given by

$$|E_1(\ell)|^2 = \frac{|E_2(0)|^2 16\pi^2\omega_1^2d_{15}^2|E_3|^2\ell^2}{c^2 (n_1)^2}, \tag{36}$$

the signal output appearing simply as mixing of the pump and idler waves.

It is shown in [2, eq. (26)] from the operator equations for a_1 and a_2^+ analogous to Eq. (35) that in the absence of an idler, i.e., $|E_2(0)|^2 = 0$, the actual signal noise output can be attributed to a (fictitious) idler energy of one photon/mode. This result is analogous to the familiar result for laser noise below threshold (fluorescence), that spontaneous emission occurs at the same rate as stimulated emission from one photon per mode [60].

The value of $|E_2(0)|^2$ associated with an energy $\hbar\omega_2$ per idler mode corresponds to an input power dP_2 given by

$$dP_2 = \frac{\hbar\omega_2 Ad\Omega_2 d\omega_2}{2\pi(\lambda_2)^2}, \tag{37}$$

where $d\Omega_2$ is the effective solid angle, and $d\omega_2$ is the frequency width of the fictitious incident power at ω_2. It is assumed that the linear medium surrounding the crystal has the same refractive index as the crystal. Since $\lambda_2 = 2\pi c/n_2\omega_2$ and $|E_2(0)|^2 = (8\pi/n_2 cA)dP_2$ we have the noise equivalent

$$|E_2(0)|^2 = \frac{\hbar n_2\omega_2^3 d\omega_2 d\Omega_2}{\pi^2 c^3}. \tag{38}$$

Since it is desirable to express dP_1 in terms of the solid angle $d\Omega_1$ defined by the aperture of the detection system, we note that $d\Omega_1/d\Omega_2 = n_2^2\omega_2^2/n_1^2\omega_1^2$ for a near-collinear experiment, and also that $d\omega_1 = d\omega_2$. Substitution of these expressions into Eq. (36) with $|E_3|^2 = (8\pi/n_3 cA)P_3$ results in

$$dP_1 = \frac{n_1}{n_2 n_3}\frac{16\hbar\omega_1^4\omega_2^2 d_{15}^2\ell^2 P_3 d\Omega_1 d\omega_1}{c^5} \tag{39}$$

Eq. (39) expresses the signal noise power dP_1 emitted in a bandwidth $d\omega_1$ and solid angle $d\Omega_1$ in terms of the pump power P_3 when the phase matching condition $\Delta k \ell \ll 1$ is satisfied over the entire range $d\Omega_1$ and $d\omega_1$. This result is in agreement with [65, eq. (25)], and is valid for small apertures and very narrow band detection. The output noise is linear in the detector solid angle and bandwidth, and is quadratic in the crystal length. The output power, unlike second harmonic generation or parametric oscillator output, is a linear function of pump power and has been shown to be essentially independent of the degree of pump coherence [63-65].

The available bandwidth is determined by the requirement that the frequency deviation from exact phase matching results in a mismatch $\Delta k \lesssim \pi/\ell$ [63-65]. Since $\left|(\partial/\partial\omega)\Delta k\right|$ is equal to $\left|v_{g1}^{-1} - v_{g2}^{-1}\right|$, the difference in inverse group velocities at ω_1 and ω_2, the effective bandwidth $d\omega_1$ for collinear noise emission is

$$\frac{d\omega_1}{2} = \frac{1}{\left|v_{g1}^{-1} - v_{g2}^{-1}\right|}\frac{\pi}{\ell} \tag{40}$$

around the phase matching frequency. The bandwidth $d\bar{\nu}_1 = (2\pi c)^{-1}d\omega_1$ is typically a few cm^{-1} for a crystal of ~1 cm length, when phase matching occurs not too close to degeneracy. It can be seen from Eq. (40) that the noise bandwidth and output power increase rapidly near degeneracy.

The total noise detected by a small aperture detector having a broad spectral band around ω_1 then becomes

$$dP_1 = \frac{n_1}{n_2 n_3}\frac{32\pi\hbar\omega_1^4\omega_2 d_{15}^2\ell P_3 d\Omega_1}{c^5\left(v_{g1}^{-1} - v_{g2}^{-1}\right)} \tag{41}$$

For broad band detection the noise is a linear function of crystal length.

Observations [54, 58] of the linear dependence of output power on crystal length and pump power are illustrated in Figs. 11 and 12. The increase is noise power near degeneracy is shown in Figure 13[54]. Tuning curves for parametric noise are similar to the oscillator tuning curves of Figs. 8-10. In the experiments reported to date observations are typically made with a detector aperture $d\Omega_1 \sim 10^{-5}$ sr, and the fractional conversion of pump to signal frequency noise is in the range 10^{-11} to 10^{-13}. The experiments are conveniently done with pump power of ~1 w.

Observations of parametric noise in solid angles $d\Omega$ of finite size require extension of the above calculations to noncollinear phase matching. The dependence of output noise frequency on observation angle for a fixed pump light direction has been reported in [55] and [56] for KDP and LiNbO$_3$ respectively. The results are generally consistent with predictions of the phase matching angle on the basis of measured refractive indices.

20

Eq. (39) expresses the signal noise power dP_1 emitted in a bandwidth $d\omega_1$ and solid angle $d\Omega_1$ in terms of the pump power P_3 when the phase matching condition $\Delta k \ell \ll 1$ is satisfied over the entire range $d\Omega_1$. and $d\omega_1$. This result is in agreement with [65, eq. (25)], and is valid for small apertures and very narrow band detection. The output noise is linear in the detector solid angle and bandwidth, and is quadratic in the crystal length. The output power, unlike second harmonic generation or parametric oscillator output, is a linear function of pump power and has been shown to be essentially independent of the degree of pump coherence [63-65].

The available bandwidth is determined by the requirement that the frequency deviation from exact phase matching results in a mismatch $\Delta k \lesssim \pi/\ell$ [63-65]. Since $|(\partial/\partial\omega)\Delta k|$ is equal to $|v_{g1}^{-1} - v_{g2}^{-1}|$, the difference in ̶p̶r̶o̶p̶a̶g̶a̶t̶i̶o̶n̶ ̶g̶r̶o̶u̶p̶ ̶v̶e̶l̶o̶c̶i̶t̶i̶e̶s̶ ̶a̶t̶ ̶ω1̶ ̶a̶n̶d̶ ̶ω2̶,̶ ̶t̶h̶e̶ ̶e̶f̶f̶e̶c̶t̶i̶v̶e̶ ̶b̶a̶n̶d̶w̶i̶d̶t̶h̶ ̶d̶ω1̶ ̶f̶o̶r̶ collinear noise emission is

$$\frac{d\omega_1}{2} = \frac{1}{|v_{g1}^{-1} - v_{g2}^{-1}|} \frac{\pi}{\ell} \tag{40}$$

around the phase matching frequency. The bandwidth $d\bar{\nu}_1 = (2\pi c)^{-1} d\omega_1$ is typically a few cm^{-1} for a crystal of ~ 1 cm length, when phase matching occurs not too close to degeneracy. It can be seen from Eq. (40) that the noise bandwidth and output power increase rapidly near degeneracy.

The total noise detected by a small aperture detector having a broad spectral band around ω_1 then becomes

$$dP_1 = \frac{n_1}{n_2 n_3} \frac{32\pi\hbar\omega_1^4\omega_2 d_{15}^2 \ell P_3 d\Omega_1}{c^5 \left(v_{g1}^{-1} - v_{g2}^{-1}\right)} \tag{41}$$

For broad band detection the noise is a linear function of crystal length.

Observations [54, 58] of the linear dependence of output power on crystal length and pump power are illustrated in Figs. 11 and 12. The increase is noise power near degeneracy is shown in Figure 13[54]. Tuning curves for parametric noise are similar to the oscillator tuning curves of Figs. 8-10. In the experiments reported to date observations are typically made with a detector aperture $d\Omega_1 \sim 10^{-5}$ sr, and the fractional conversion of pump to signal frequency noise is in the range 10^{-11} to 10^{-13}. The experiments are conveniently done with pump power of ~ 1 w.

Observations of parametric noise in solid angles $d\Omega$ of finite size require extension of the above calculations to noncollinear phase matching. The dependence of output noise frequency on observation angle for a fixed pump light direction has been reported in [55] and [56] for KDP and LiNbO$_3$ respectively. The results are generally consistent with predictions of the phase matching angle on the basis of measured refractive indices.

aperture area A, with the pump light propagating along $\hat{x}$. Under phase matched conditions $\Delta k \ell \ll 1$, the classical Eqs. (18) for the amplified signal and idler fields $E_1(x)$ and E_2^* become

$$E_1(x) = E_1(0)\cosh sx + i\beta E_2^*(0)\sinh sx$$

$$E_2^*(x) = E_2^*(0)\cosh sx - i\beta^{-1}E_1(0)\sinh sx \qquad (35)$$

where $\beta = (E_3 n_2\omega_1/E_3^*n_1\omega_2)^{\frac{1}{2}}$, E_3 is the pump field amplitude, s is given by Eqs. (19) and (20), and n_1 and n_2 are the refractive indices at ω_1 and ω_2 respectively.

At low gain, $sx \ll 1$, and in the absence of an input signal, $E_1(0) = 0$, it follows from Eq. (35) that the signal output at $x = \ell$ is given by

$$|E_1(\ell)|^2 = \frac{|E_2(0)|^2 \, 16\pi^2\omega_1^2 d_{15}^2 \, |E_3|^2\ell^2}{c^2\,(n_1)^2}, \qquad (36)$$

the signal output appearing simply as mixing of the pump and idler waves.

It is shown in [2, eq. (26)] from the operator equations for a_1 and a_2^+ analogous to Eq. (35) that in the absence of an idler, i.e., $|E_2(0)|^2 = 0$, the actual signal noise output can be attributed to a (fictitious) idler energy of one photon/mode. This result is analogous to the familiar result for laser noise below threshold (fluorescence), that spontaneous emission occurs at the same rate as stimulated emission from one photon per mode [60].

The value of $|E_2(0)|^2$ associated with an energy $\hbar\omega_2$ per idler mode corresponds to an input power dP_2 given by

$$dP_2 = \frac{\hbar\omega_2 A d\Omega_2 d\omega_2}{2\pi(\lambda_2)^2}, \qquad (37)$$

where $d\Omega_2$ is the effective solid angle, and $d\omega_2$ is the frequency width of the fictitious incident power at ω_2. It is assumed that the linear medium surrounding the crystal has the same refractive index as the crystal. Since $\lambda_2 = 2\pi c/n_2\omega_2$ and $|E_2(0)|^2 = (8\pi/n_2 cA)dP_2$ we have the noise equivalent

$$|E_2(0)|^2 = \frac{\hbar n_2\omega_2^3 \, d\omega_2 d\Omega_2}{\pi^2 c^3}. \qquad (38)$$

Since it is desirable to express dP_1 in terms of the solid angle $d\Omega_1$ defined by the aperture of the detection system, we note that $d\Omega_1/d\Omega_2 = n_2^2\omega_2^2/n_1^2\omega_1^2$ for a near-collinear experiment, and also that $d\omega_1 = d\omega_2$. Substitution of these expressions into Eq. (36) with $|E_3|^2 = (8\pi/n_3 cA)P_3$ results in

$$dP_1 = \frac{n_1}{n_2 n_3} \frac{16\hbar\omega_1^4\omega_2 d_{15}^2 \ell^2 P_3 d\Omega_1 d\omega_1}{c^5} \qquad (39)$$

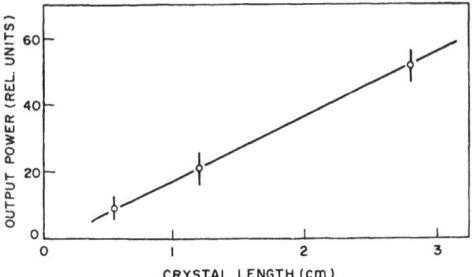

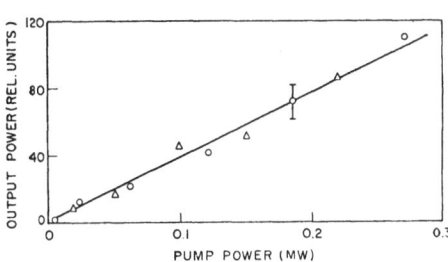

FIGURE 11. Optical parametric noise power in ADP as a function of crystal length (after Magde and Mahr, [54]). A linear dependence is expected in this experiment for the reasons discussed in the text, and also because the coherence length is limited to 0.3 cm by the pump beam angular divergence. The noise power is observed at 7800 Å; the pump is at 3472 Å.

FIGURE 12. Parametric noise in ADP as a function of pump power. The symbols indicate different crystal areas (after Magde and Mahr, [54]). The figure illustrates the linear dependence on pump power, and the lack of dependence on pump intensity.

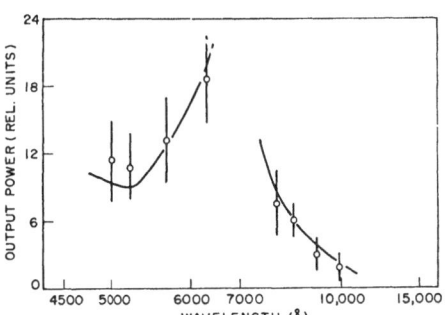

FIGURE 13. Parametric noise in ADP as a function of wavelength (after Magde and Mahr, [54]). The peak near the degenerate wavelength (6943 Å) arises from the increase in noise bandwidth indicated by Eq. (40).

The theory of optical parametric noise viewed as a three-photon scattering process has been discussed [62-65]. It has been shown [64, 65] for the case of noncollinear emission that at angles of maximum deviation from the input pump light consistent with phase matching noise is emitted with substantially greater than the usual bandwidth and power. This enhancement in signal frequency noise emission, for example, can be looked at as due to a singularity in the number of idler frequency modes contributing to the signal frequency noise in a given solid angle $d\Omega_1$. Noise power in these unique directions is predicted to be proportional to $\ell^{3/2}$ rather than ℓ.

Residual optical parametric noise is also expected in directions for which phase matching is impossible[64, 65]. The nonphase matched noise intensity is much smaller than the phase matched value and has not yet been observed. Its frequency distribution is predicted to show singularities at those limiting frequencies beyond which t r a n s v e r s e momentum conservation is impossible.

Phase matched parametric noise shows promise of providing a useful tool for the measurement of nonlinear optical coefficients, since in a well designed experiment the noise intensity is highly insensitive to the details of the spatial and temporal coherence of the incoming light.

IV. THE STIMULATED RAMAN EFFECT

We now discuss the amplification of Stokes light in the stimulated Raman effect. It is well known that this process is an example of a parametric interaction of the laser pump field (ω_p), the Raman Stokes field (ω_s), and the coherent molecular vibrations or optical phonons (ω_0). The parametric character becomes most evident when the interaction is written in coupled wave form [10, 66, 67]. As is clear from the papers by Ducuing[9] and Maier[67], the main features of the stimulated Raman effect can be described classically. The justification for the use of classical fields, and the sense in which the quantum field representation approaches classical behavior for intense fields have recently been discussed[68, 69].

We describe here first the role of the molecular population distribution in the conventional coupled wave formulation of stimulated Raman gain, showing that the resonant molecular response can be represented vectorially in a way similar to the resonant response of an electric or magnetic dipole system to electromagnetic waves[70]. Secondly, we show that in backward stimulated Raman emission[67, 71, 72] the amplified Stokes pulse is expected to approach a limiting shape and energy similar to the limiting pulses expected in laser amplification[73] and self-induced transparency[74].

Consider stimulated Raman scattering in CS_2 due to the infrared inactive stretching vibration at $\bar{\nu} = \omega_0 / 2\pi c = 656$ cm^{-1}. Molecular anharmonicity in CS_2 introduces splitting between the $(100) \leftarrow (000)$ and $(200) \leftarrow (100)$ transitions which is large compared to the line width. As a result, the $(200) \leftarrow (100)$ transition can be ignored to a good approximation in discussing resonant interactions at 656 cm^{-1}. We therefore represent the molecules as an ensemble of randomly oriented two level systems, the effective electronic polarizability of each molecule having a linear dependence on the vibrational coordinate q given by $\partial\alpha/\partial q$. The molecular hamiltonian is

$$\mathcal{H}' = \mathcal{H}^0 - \frac{1}{2}\frac{\partial\alpha}{\partial q} q' E^2 \tag{42}$$

in the presence of optical electric field E. Let a_0 and a_1 represent the amplitudes of the (000) and (100) states in the Schroedinger representation. It is useful to define $\delta \equiv a_0 a_0^* - a_1 a_1^*$, the difference in probability of finding a molecule in the ground and excited states, and $\tilde{q} \equiv \langle q' \rangle$, the expectation value of the operation q'. From Eq. (42) and the Heisenberg equation of motion one obtains[75] the equations:

$$\frac{\partial^2\tilde{q}}{\partial t^2} + \Gamma\frac{\partial\tilde{q}}{\partial t} + \omega_0^2\,\tilde{q} = \frac{1}{2m}\frac{\partial\alpha}{\partial q}E^2\,\delta$$

22

$$P = N \frac{\partial \alpha}{\partial q} \tilde{q} \, E$$

$$\frac{\partial \delta}{\partial t} = -\frac{1}{\hbar \omega_o} \left(\frac{\partial \alpha}{\partial q} \right) E^2 \frac{\partial \tilde{q}}{\partial t} + \Gamma' \, (1 - \delta)$$

$$\frac{\partial^2 E}{\partial z^2} - \frac{1}{c^2} \frac{\partial^2 D}{\partial t^2} = \frac{4\pi}{c^2} \frac{\partial^2 P}{\partial t^2} . \tag{43}$$

In Eqs. (43), $m \equiv \hbar (2\omega_o q_{01}{}^2)^{-1}$, N is the molecular density, P is the nonlinear polarization, and D the displacement vector including linear polarization. The phenomenological damping constants Γ and Γ' describe the dephasing rate due to collisions ($\Gamma \cdot T_2^{-1}$) and the population relaxation rate ($\Gamma' \sim T_1^{-1}$) respectively. The full width at half intensity of the Raman line is equal to Γ. The small Boltzmann population of the (100) state is neglected here, and δ is taken to have an equilibrium value of unity.

The optical electric field considered here consists of laser pump (ω_p) and Raman Stokes (ω_s) plane wave components of the form

$$E = E_s \cos (\omega_s t - k_s z) + E_p \cos (\omega_p t + k_p z), \tag{44}$$

propagating in opposite directions. We consider the resonant case

$$\omega_s = \omega_p - \omega_o. \tag{45}$$

The amplitudes E_s and E_p are slowly varying functions of z and t. A significant interaction[72] occurs only with a molecular vibrational wave of the form

$$\tilde{q} = q \cos (\omega_o t + k_o z - \pi/2) \tag{46}$$

where

$$k_o = k_p - k_s \tag{47}$$

and q is a slowly varying function of z and t. The three fields satisfy the phase relation $\psi_p - (\psi_s + \psi_o) = \pi/2$ of Eq. (1) for parametric gain. Substitution of Eqs. (44) and (46) into Eqs. (43) leads to the coupled amplitude equations

$$\frac{\partial q}{\partial t} + \frac{\Gamma}{2} q = - \frac{1}{4m\omega_o} \frac{\partial \alpha}{\partial q} E_p E_s \, \delta$$

$$\frac{\partial \delta}{\partial t} = \frac{1}{2\hbar} \frac{\partial \alpha}{\partial q} E_p E_s \, q + \Gamma' \, (1 - \delta)$$

$$\frac{\partial E_s}{\partial z} + \frac{n_s}{c} \frac{\partial E}{\partial t} + \frac{\gamma_s}{2} E_s = \frac{-\pi}{n_s c} \frac{N \omega_s}{c} \frac{\partial \alpha}{\partial q} E_p q$$

$$\frac{\partial E_p}{\partial z} - \frac{n_p}{c} \frac{\partial E_p}{\partial t} - \frac{\gamma_p}{2} E_p = \frac{-\pi N \omega_p}{n_p c} \frac{\partial \alpha}{\partial q} E_s q \qquad (48)$$

These semiclassical equations are similar to the classical eqs. (6)-(8) of Maier's paper [67] but include and describe modifications due to changes in vibrational state population caused by the stimulated Raman generation. Eqs. (48) include linear losses γ_s and γ_p at the Stokes and the laser pump frequencies. These equations approach the classical coupled wave equations when $\delta \to 1$, the limit of negligible population change and a good approximation in the case of the liquid CS_2 experiments of Maier et al[67].

As discussed by Maier [67] Eqs. (48) lead in the limit $\delta \to 1$, $\frac{\partial q}{\partial t} \ll \frac{\Gamma}{2} q$ to rate equations which describe the growth of backward Raman Stokes pulses. These pulses have been observed[71] to have peak power an order of magnitude higher than the incident laser pump power and durations as short as 20 psec; their properties are described in detail elsewhere[72].

As the backward pulse duration becomes comparable with $T_2 \sim 1/\Gamma$, the amplification needs to be described by Eqs. (48) rather than by the rate equations[67]. We discuss here some aspects of the solutions of Eqs. (48) when the pulse duration is short relative to T_2, i.e. $\frac{\partial q}{\partial t} \gg \frac{\Gamma}{2} q$.

When dephasing and relaxation can be neglected, the molecular response is similar to the resonant response of electric or magnetic dipoles to the electromagnetic field. In the limit $\Gamma \to 0$, $\Gamma' \to 0$ it follows from Eqs. (48) that

$$\frac{2m\omega_0}{\hbar} q^2 + \delta^2$$

is a constant. It is thus possible to define a unit state vector $\underset{\sim}{V}$ having vertical component δ and horizontal component $(2m\omega_0/\hbar)^{\frac{1}{2}} q$, with direction angle ε (Figure 14). The effect of the fields E_p and E_s is to cause $\underset{\sim}{V}$ to precess about an axis normal to the page at an angular frequency

$$\frac{d\varepsilon}{dt} = \frac{1}{(2m\omega_0 \hbar)^{\frac{1}{2}}} \left(\frac{\partial \alpha}{\partial q}\right) \frac{E_p E_s}{2}. \qquad (49)$$

This precession is analogous to the resonant precession of an electric dipole moment μ at angular frequency $\mu E/\hbar$ in an electric field E. The vector model is of course applicable generally to the stimulated Raman effect, and is not limited to the backward case discussed here. The vector model of stimulated Raman processes has been developed independently by Ducuing [9].

An interesting property of backward stimulated Raman scattering is the existence of a steady state Stokes pulse in the limit $\frac{\partial q}{\partial t} \gg \frac{\Gamma}{2} q$[72]. We look for a steady state solution satisfying

$$\frac{\partial E_s}{\partial z} + \frac{n_s}{c} \frac{\partial E_s}{\partial t} = 0$$

$$\frac{\partial E_p}{\partial t} + \frac{n_s}{c} \frac{\partial E_p}{\partial t} = 0 \tag{50}$$

as illustrated in Figure 15. Eqs. (48) and (50) have a useful analytic solution in the limit in which the variation of δ can be neglected.

As a boundary condition we take the incident pump photon flux to be a constant, N_p cm^{-2} sec^{-1}; let $N_s(t)$ represent the instantaneous Stokes photon flux. The initial value of q is taken to be small. It follows that a steady state pulse exists, of the form [12]

$$\frac{N_s}{N_p} = 1.76 \frac{n}{n_s} \frac{\tau_s}{\tau_{\frac{1}{2}}} \operatorname{sech}^2 \left(\frac{1.76t}{\tau_{\frac{1}{2}}}\right) \tag{51}$$

where

$$\frac{1.76}{\tau_{\frac{1}{2}}} = \frac{\Gamma}{2} \frac{\gamma_g}{\gamma_s} . \tag{52}$$

The average refractive index $n \equiv (n_s + n_p)/2$, the damping time for Stokes light due to the linear losses is $\tau_s = n_s(\gamma_s c)^{-1}$, and the width $\tau_{\frac{1}{2}}$ of the steady state pulse at half intensity is given by Eq. (52). The Stokes gain γ_g is the small signal gain (nepers cm^{-1}) calculated from Eqs. (48) in the rate equation approximation. The time scale of Eq. (51) is such that the pulse maximum passes the observation point at t = 0.

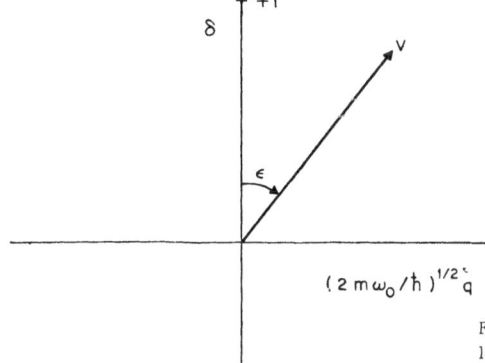

FIGURE 14. Unit vector $\underset{\sim}{V}$ describing state of "2 level" Raman active vibrational transition. δ is the difference in probability of finding the molecule in the upper and lower states; q is the expectation value of the vibrational displacement. $\underset{\sim}{V}$ precesses due to the combined action of the laser and Raman Stokes fields.

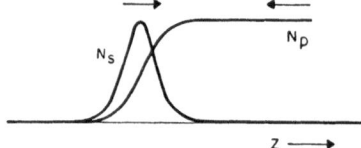

$z \longrightarrow$

FIGURE 15. Schematic diagram of the steady state backward stimulated Raman Stokes pulse. N_s and N_p are the photon flux densities at the Stokes and laser pump frequencies. The pulse duration $\tau_{\frac{1}{2}}$ (tan)$<<$ Γ^{-1}, the full width of the Raman line.

The steady state Raman pulse of Eq. (51) has the following interpretation. As the backward traveling pulse is amplified and narrowed in the region $\tau_{\frac{1}{2}} << \Gamma^{-1}$, its spectral width becomes correspondingly broader relative to Γ. The ratio of the pulse spectral width to the Raman line width is $\sim \Gamma \tau_{\frac{1}{2}})^{-1}$. The effective gain seen by the pulse is expected to be reduced from its steady state value by just this ratio, since only a fraction $\sim \Gamma \tau_{\frac{1}{2}}$ of the pulse energy can effectively interact with the Raman line. A steady state condition is reached when the effective gain becomes just equal to the linear loss γ_g. It is this condition that is expressed by Eq. (52). The steady state backward Raman pulse is analogous to the steady state laser pulse discussed in [73] for the case of an amplifying medium with residual linear loss. It differs however in not representing a "180° pulse" with respect to the vibrational population, which in the present approximation is hardly changed. The pulse does however completely deplete the incident pump beam. As the pulse propagates through the incident pump light the latter is completely transformed to Stokes light and absorbed via the linear loss mechanism. It is of interest that the (sech)^2 pulse shape is identical to that predicted for laser amplification [73] and self-induced transparency [74].

It will be apparent that this mechanism of short pulse formation is not restricted to the stimulated Raman effect, but should be observable in any backward wave parametric amplifier[52, 76] of adequate gain.

The examples discussed in Sections II to IV are but a few aspects of parametric processes recently encountered in optics. It can confidently be predicted that many new examples of parametric optical phenomena remain to be discovered.

REFERENCES

1. W. H. LOUISELL, Coupled Modes and Parametric Electronics, John Wiley and Sons, Inc., New York (1969)
2. W. H. LOUISELL, A. YARIV, and A. E. SIEGMAN, Phys. Rev., 124, 1646 (1961)
3. B. R. MOLLOW and R. J. GLAUBER, Phys. Rev., 160, 1076 (1967)
4. B. R. MOLLOW and R. J. GLAUBER, Phys. Rev., 160, 1097 (1967)
5. R. J. GLAUBER, in Quantum Optics and Electronics, Ed. C. de Witt et al. Gordon and Breach, New York, p. 63 (1964)
6. W. W. MUMFORD, Proc. Instn. Radio Engrs., 48, 848 (1960)
7. H. SUHL, J. Appl. Phys., 28, 1225 (1957)
8. P. A. FRANKEN, A. E. HILL, C. W. PETERS, and G. WEINREICH, Phys. Rev. Letters, 7, 118 (1961)

9. J. DUCUING, Proc. of the International Summer School of Physics, Enrico Fermi, Course 42, Quantum Optics, ed. by R. J. G. Glauber, Academic Press, New York (1969)

10. N. BLOEMBERGEN, Nonlinear Optics, W. A. Benjamin, Inc., New York (1965)

11. J. A. ARMSTRONG, N. BLOEMBERGEN, J. DUCUING, and P. S. PERSHAN, Phys. Rev., **127**, 1918 (1962)

12. P. S. PERSHAN, Phys. Rev., **130**, 919 (1963)

13. G. D. BOYD and D. A. KLEINMAN, J. Appl. Phys. (to be published). A useful discussion of the definition of the nonlinear coefficients is given in Appendix 2

14. J. F. NYE, Physical Properties of Crystals, Oxford (1960)

15. D. A. KLEINMAN, Phys. Rev., **126**, 1977 (1962)

16. P. L. KELLEY, J. Chem. Phys. Solids, **24**, 607 (1963)

17. P. J. PRICE, Phys. Rev., **130**, 1792 (1963)

18. H. CHENG and P. B. MILLER, Phys. Rev., **134**, A683 (1964)

19. J. F. WARD, Rev. Mod. Phys., **37**, 1 (1965)

20. P. A. FRANKEN and J. F. WARD, Rev. Mod. Phys., **35**, 23 (1963)

21. C. G. B. GARRETT and F. N. H. ROBINSON, IEEE J. Quantum Elec., Qe-2, 328 (1966)

22. F. N. H. ROBINSON, Bell System tech. J., **46**, 913 (1967)

23. J. A. GIORDMAINE, Phys. Rev., **138**, A1599 (1965)

24. R. C. MILLER, Appl. Phys. Letters, **5**, 17 (1964)

25. P. P. BEY, J. F. GIULIANI, and H. RABIN, Phys. Rev. Letters, **19**, 819 (1967)

26. J. A. GIORDMAINE, Phys. Rev. Letters, **8**, 19 (1962)

27. P. D. MAKER, R. W. TERHUNE, M. NISENHOFF, and C. M. SAVAGE, Phys. Rev. Letters, **8**, 21 (1962)

28. R. C. MILLER, G. D. BOYD, and A. SAVAGE, Appl. Phys. Letters, **6**, 77 (1965)

29. A. ASHKIN, G. D. BOYD, J. M. DZIEDZIC, R. G. SMITH, A. A. BALLMAN, H. J. LEVINSTEIN, and K. NASSAU, Appl. Phys. Letters, **9**, 72 (1966)

30. F. S. CHEN, J. Appl. Phys., **38**, 3418 (1967)

31. L. G. VAN UITERT, S. SINGH, H. J. LEVINSTEIN, J. E. GEUSIC and W. A BONNER, Appl. Phys. Letters, **11**, 161 (1967), **12**, 224 (1968)

32. J. E. GEUSIC, H. J. LEVINSTEIN, J. J. RUBIN, S. SINGH, and L. G. VAN UITERT, Appl. Phys. Letters, **11**, 269 (1967), **12**, 224 (1968)

33. J. A. GIORDMAINE and R. C. MILLER, Phys. Rev. Letters, **14**, 973 (1965)

34. J. A. GIORDMAINE and R. C. MILLER, in Physics of Quantum Electronics, Ed. P. L. Kelley et al., McGraw-Hill Book Co., New York, p. 31 (1966)

35. The corresponding expressions for P used in [33] and [34] are too small by a factor of 2, and the expressions for $\gamma_0^{[33]}$ and $\gamma^{[34]}$ should therefore be multiplied by 2. Since the value d_{15} quoted in [33] and [34] is now known to be too large by a factor of 2.6 (Table I), the calculated threshold power in [33] and [34] is approximately correct.

36. A. YARIV and W. H. LOUISELL, IEEE J. Quantum Elect., QE-2, 418 (1966)

37. G. D. BOYD and A. ASHKIN, Phys. Rev., **146**, 187 (1966)

38. R. C. MILLER and W. A. NORDLAND, Appl. Phys. Letters, **10**, 53 (1967)

39. J. A. GIORDMAINE and R. C. MILLER, Appl. Phys. Letters, **9**, 298 (1966)

40. S. A. AKHMANOV, A. I. KOVRIGIN, V. A. KOLOSOV, A. S. PISKARSKAS, V. V. FADEEV, and R. V. KHOKHLOV, ZhETF Pis'ma, **3**, 372 (1966); Soviet Phys. JETP Letters, **3**, 241 (1966)

41. L. B. KREUZER, Appl. Phys. Letters, **10**, 336 (1967)

42. R. H. KINGSTON, Proc. Instn. Radio Engrs., **50**, 472 (1962)

43. N. M. KROLL, Phys. Rev., **127**, 1207 (1962)

44. S. A. AHKMANOV and R. V. KHOKHLOV, ZhETF, **43**, 351 (1962); Soviet Phys. JETP, **16**, 252 (1963)

45. J. A. GIORDMAINE and D. A. KLEINMAN, Efficient Optical Harmonic Generation, Parametric Amplification, Oscillation and Modulation, U.S. Patent 3,234,475, Feb. 8, 1966; (filed Dec. 11, 1961)

46. S. A. AKHMANOV and R. V. KHOKHLOV, Problemy nelineinoi optiki (Problems of Nonlinear Optics), VINITI, Moscow (1964); Usp. Fiz. Nauk, **88**, 439 (1966); Soviet Phys. Uspekhi, **9, 210** (1966)

27

47. S. A. AKHMANOV, V. G. DMITRIYEV, V. P. MODENOV, and V. V. FADEYEV, Radiotekhnika i Elektronika, **10**, 649 (1965); Rad. Engng. electron. Phys., **12**, 1841 (1965)

48. Y. V. GRIGOREV, V. K. RUBENKO, and R. V. KHOKHLOV, Izv. Vyssh. Ucheb. Zaved. Radiofiz., **9**, 932 (1966)

49. R. H. KINGSTON and A. L. McWHORTER, Proc. IEEE, **53**, 4 (1965)

50. A. YARIV, IEEE J. Quantum Elect., QE-2, 30 (1966)

51. S. E. HARRIS, IEEE J. Quantum Elect., QE-2, 701 (1966)

52. S. E. HARRIS, Appl. Phys. Letters **9**, 114 (1966)

53. S. E. HARRIS, M. K. OSHMAN, and R. L. BYER, Phys. Rev. Letters, **18**, 732 (1967)

54. D. MAGDE and H. MAHR, Phys. Rev. Letters, **18**, 905 (1967)

55. S. A. AKHMANOV, A. G. ERSHOV, V. V. FADEYEV, R. V. KHOKHLOV, O. N. CHUNAEV, and E. V. SHVOM, Zh ETF Pis'ma **2**, 458 (1965); Soviet Phys. JETP Letters **2**, 285 (1965)

56. R. G. SMITH, J. G. SKINNER, J. E. GEUSIC and W. G. NILSEN, Appl. Phys. Letters **12**, 97 (1968)

57. R. L. BYER and S. E. HARRIS, Phys. Rev., **168**, 1064 (1968)

58. J. A. GIORDMAINE and E. C. MILLER (unpublished)

59. J. P. GORDON, W. H. LOUISELL, and L. R. WALKER, Phys. Rev. Letters, **129**, 481 (1963)

60. W. H. LOUISELL, Radiation and Noise in Quantum Electronics, McGraw-Hill Book Co., New York (1964)

61. W. G. WAGNER and R. W. HELLWARTH, Phys. Rev., **133**, A915 (1964)

62. D. N. KLYSHKO, Zh ETF Pis'ma, **6**, 490 (1967); Soviet Phys. JETP Letters, **6**, 23 (1967)

63. T. G. GIALLORENZI and C. L. TANG, Phys. Rev., **166**, 225 (1968)

64. D. A. KLEINMAN, Bul. Am. Phys. Soc., **13**, 439 (1968)

65. D. A. KLEINMAN, Phys. Rev., **174**, 1027 (1968)

66. A Review of the stimulated Raman effect is given in N. BLOEMBERGEN, Am. J. Phys., 35, 989 (1967)

67. M. MAIER, W. KAISER, M. STANKA, and J. A. GIORDMAINE, Proc. of the International Summer School of Physics, Enrico Fermi, Course 42, Quantum Optics, ed., by R. J. Glauber, (Academic Press, New York, 1969)

68. Y. R. SHEN, Phys. Rev., **155**, 921 (1967)

69. H. S. FREEDHOFF, J. Chem. Phys., **47**, 2554 (1967)

70. E. T. JAYNES and F. W. CUMMINGS, Proc. IEEE, **51**, 89 (1963); R. P. FEYNMAN, F. L. VERNON, Jr., and R. W. HELLWARTH, J. Appl. Phys., **28**, 49 (1957)

71. M. MAIER, W. KAISER, and J. A. GIORDMAINE, Phys. Rev. Letters, **17**, 1275 (1966)

72. M. MAIER, W. KAISER, J. A. GIORDMAINE, Phys. Rev., **177**, 580 (1969)

73. F. T. ARECCHI and R. BONIFACIO, IEEE J. Quantum Elect., Q. E. -1, 169 (1965)

74. S. L. McCALL and E. L. HAHN, Phys. Rev. Letters, 18, 908 (1967); V. E. KHARTSIEV, Physics, **3**, 129 (1967)

75. J. A. GIORDMAINE and W. KAISER, Phys. Rev., **144**, 676 (1966) In eqs. (2) and (3) and subsequent equations, H' and F should be multiplied by $\frac{1}{2}$. Eq. (9) should read, $\hbar\omega_0(d\delta\ dt) = -2F(dq/dt)$.

76. H. HSU and K. F. TITTLE, in Lasers and Applications, Ed. by W. S. C. Chang, Ohio State Univ., p. 192 (1963)

2/Holography and its Applications

ALBERT A. FRIESEM

INTRODUCTION

Wavefront reconstruction, generally called holography, represents a
revolutionary photographic process. In holography, both the amplitude and
phase of the wavefronts scattered by an object are recorded by interferometric methods. At some arbitrary later time, the original wave pattern
is recreated by illuminating the recorded photograph, called a hologram,
with a coherent beam of light. The object scene is thereby revealed in
full three dimension, possessing the visible properties of parallax between
near and far objects, and perspective that changes with the viewer's position.
Numerous applications are currently being investigated in laboratories
throughout the United States. The major aim of this paper is to discuss
the principles of holography and to describe some of the important
applications.

BASIC ANALYSIS

The basic process of hologram formation is described with the aid of
Figure 1. Shown here are two waves, the object wave o and the reference
wave r, which is coherent with
o. The latter aids in capturing
the phase of the signal bearing
wave o; this operation is the
basic property which distinguishes
holography from conventional
photography. The total field
impinging on the photosensitive
medium is then r + o.

Since the photosensitive
medium responds to the intensity
of the incident radiation, the
desired information which becomes
the resultant amplitude transmittance of the medium is represented
as

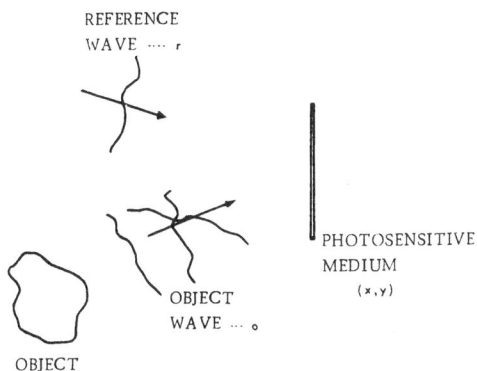

FIGURE 1. Basic process of holographic recording.

$$T_a(x, y) = g[I(x,y)] = g[(r + o)(r + o)^*] \tag{1}$$

which is, in general, a nonlinear relation. Over a linear range of transmittance, a reasonable approximation of the actual transmission versus exposure characteristics is given by

$$T_a = T_o - k\ I = T_o - k[rr^* + oo^* + ro^* + r^*o] \tag{2}$$

where I is the intensity distribution across the film and T_o and k are constants determined by the recording medium characteristics. Eq. (2) is a mathematical description of the amplitude transmittance of the hologram. It should be noted that both the amplitude and phase of the wavefront o have thus been recorded.

If the hologram is illuminated with a "reconstruction" wavefront, c, the emerging wave field has the form cT_a. In its expanded form, it becomes

$$cT_a = cT_o - kcrr^* - kcoo^* - kcro^* - kcr^*o \tag{3}$$

where the final term in Eq. (3) represents a wave that is a replica of the original object wave, possessing all its properties, and capable of forming an image of the original object. The implication here is that this part of the emerging wavefront can be separated from the other parts so as not to get any overlap. For some time this was a major obstacle in holography, and although many schemes have been proposed to implement separation, the most successful is the off-axis reference beam or spatial carrier frequency method.

Because of this separation, and since the mapping from input-to-output signal is linear, the important superposition principle of linear systems may be employed. We therefore consider point source object for analyzing hologram behavior; the results for a more general object can be found by superposition. In this discussion, we shall only consider first order analysis.

The recording and reconstructing geometries are shown in Figure 2. The illumination wavelength during recording is λ_1, and wavelength for reconstructing is λ_2. At the recording plane, the contribution of the various waves are

object wave $\qquad\qquad o = A_o(x, y)\ e^{j\phi_o(x, y)}$

reference wave $\qquad\qquad r = A_r(x, y)\ e^{j\phi_r(x, y)} \qquad (4)$

reconstructing wave $\qquad c = A_c(x, y)\ e^{j\phi_c(x, y)}$

Let each phase term of the three waves be approximated by

$$\phi_m(x_1 y) = \frac{2\pi}{\lambda_m}\left[\ \frac{1}{2\,z_m}\ (x^2 + y^2 - 2xx_m - 2yy_m + \ldots)\right] \tag{5}$$

where the subscript m denotes any of the three subscripts o, r, c.

30

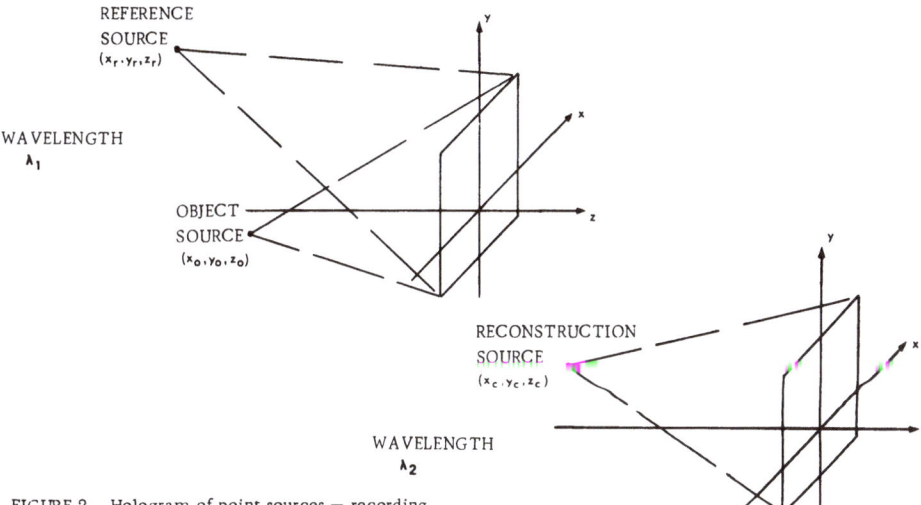

FIGURE 2. Hologram of point sources — recording and reconstruction geometries.

a) recording geometry
b) reconstruction geometry

Substitution of Eqs. (4) and (5) into the last two terms of Eq. (3) results in

$$kcro*, \ kcr*o = kA_0A_rA_c \ exp \ j\pi \left\{ \left(\pm \frac{1}{\lambda_1 z_r} \mp \frac{1}{\lambda_1 z_0} + \frac{1}{\lambda_2 z_c} \right)(x^2+y^2) \right.$$

$$+ \left(\pm \frac{x_0}{\lambda_1 z_0} - \frac{x_c}{\lambda_2 z_c} \mp \frac{x_r}{\lambda_1 z_r} \right) x$$

$$\left. + \left(\pm \frac{y_0}{\lambda_1 z_0} - \frac{y_c}{\lambda_2 z_c} \mp \frac{y_r}{\lambda_1 z_r} \right) y \right\} \tag{6}$$

where K is a positive constant.

Expression (6) leads to the following image location

$$z_i = \frac{\lambda_1 \ z_r z_c z_0}{\lambda_1 z_0 z_r \pm \lambda_2 z_0 z_c \mp \lambda_2 z_c z_r}$$

$$x_i = \frac{\lambda_1 \ z_r z_0 x_c \pm \lambda_2 z_0 z_c x_r \mp \lambda_2 \ z_r z_c x_0}{\lambda_1 \ z_r z_0 \pm \lambda_2 z_0 z_c \mp \lambda_2 z_c z_r} \tag{7}$$

$$y_i = \frac{\lambda_1 \ z_r z_0 y_c \pm \lambda_2 \ z_0 z_c y_r \mp \lambda_2 z_r z_0 y_0}{\lambda_1 \ z_r z_0 \pm \lambda_2 \ z_0 z_c \mp \lambda_2 z_c z_r}$$

where positive z_i denotes virtual image and negative z_i the real image. The lateral and longitudinal magnifications for the reconstructed image point are

$$M_{lat} = \left|\frac{\Delta x_i}{\Delta x_0}\right| = \left|\frac{\Delta y_i}{\Delta y_0}\right| = \left|\frac{\lambda_2 z_i}{\lambda_1 z_0}\right| = \left|1 - \frac{z_0}{z_r} \mp \frac{\lambda_1 z_0}{\lambda_2 z_c}\right|^{-1}$$

and   (8)

$$M_{long} = \left|\frac{\Delta z_i}{\Delta z_0}\right| = \left|\frac{\lambda_2 z_i^2}{\lambda_i z_0^2}\right| = \frac{\lambda_1}{\lambda_2} M_{lat}^2$$

Eq. (8) demonstrates the magnification possibilities with holograms which may be achieved either by choosing different geometrical source positions and/or different wavelengths in recording and reconstruction. They also demonstrate that nonuniform magnification of three-dimensional objects will take place. The nonuniformity may be overcome by scaling the hologram in the same ratio as the wavelength; this operation, unfortunately, is not readily achieved and is fraught with many practical difficulties.

BASIC SCHEME

The basic scheme of constructing a hologram is shown in Figure 3. The incident coherent light is split into two beams. The signal beam is that part of the light which is scattered by the object and exposes the photographic plate. The reference beam is directed by means of a mirror onto the photographic plate where it interferes with the signal beam. The resulting recorded complex interference pattern is a hologram. The reference wave is conventionally a plane wave or a spherical wave, but it may be any wavefront shape which can be conveniently duplicated when subsequently viewing the hologram.

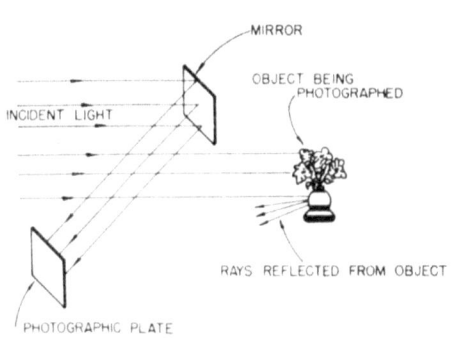

MAKING OF HOLOGRAM

FIGURE 3. System for holographic recording of three dimensional reflecting objects (courtesy of Leith and Upatnieks).

The reconstructed images are readily viewed through the hologram which has been placed in a beam of coherent light. This is schematically shown in Figure 4. The virtual image is seen by looking into the hologram, as if it were a window, and is indistinguishable from the original object. The real image forms in front of the hologram and may be seen by an observer as an image suspended in space between himself and the hologram. The real image is unfortunately pseudoscopic which causes difficulty in observation.

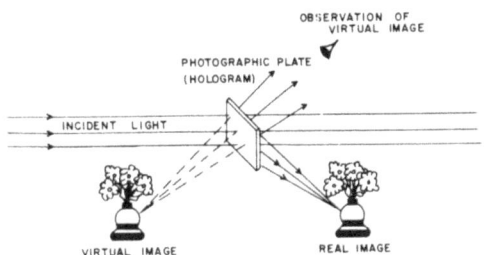

FIGURE 4. Reconstruction from a hologram showing
real and virtual images (courtesy of Leith and Upatnieks).

RECORDING MEDIUM TRANSFER CHARACTERISTICS

The recording medium, typically photographic film or photographic plates,
serves an important role in holography. It is used as a storage device,
recording the fluctuations in intensity which result from the interference
of the reference waves with the signal waves. After appropriate development, this record is interrogated with a coherent light beam providing
a reconstructed image.

A most suitable characterization of the photographic film is in terms
of its amplitude transmittance versus its exposure. A representative
sample and the utilization of these transfer characteristics is illustrated
in Figure 5. The sinusoidal function plotted along the vertical axis is the
spatial variation of exposure, and the sinusoid plotted along the
horizontal axis is the spatial variation of amplitude transmittance. Linear
recording is achieved when a sinusoidal exposure produces an undistorted,
sinusoidal, amplitude transmittance variation.

For the case of a simple grating hologram, which results from the
recording of two plane waves, the amplitude transmittance is written as

$$T_a = T_b + 2\beta (I_r I_0)^{\frac{1}{2}} \left[tK\left(\frac{1}{\alpha}\right) \right] \cos \left[\frac{2\pi}{\alpha} x - \phi\left(\frac{1}{\alpha}\right) \right] \qquad (9)$$

where
I_0 is the intensity of signal beam,
I_r is the intensity of reference beam,
T_b is the amplitude transmittance bias caused by the spatially constant
     intensity I_r and I_0,
β is the slope of the curve of the amplitude transmittance versus
     exposure at the bias point,
α is the period of the grating fringes in the x-direction,
t   is the exposure time, and
K and ϕ are dependent on the spatial frequency and define an optical
     transfer function of the film.

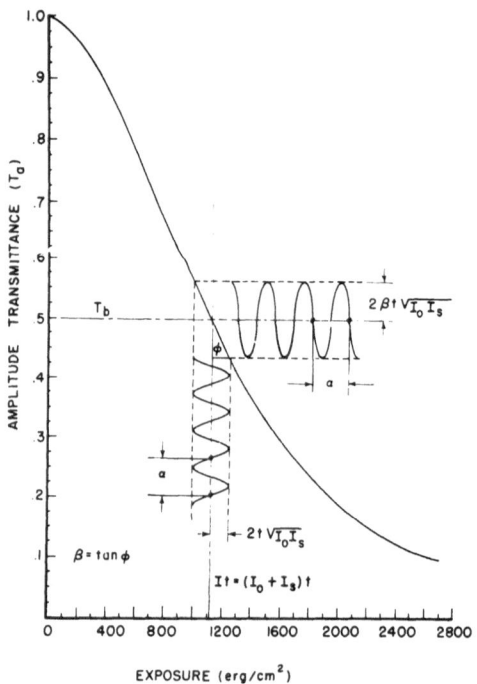

FIGURE 5. Transfer characteristics; Kodak 649F spectroscopic plates.

We can readily choose the operating parameters for the recording process. Exposure time adjustment determines the operating point T_b of Figure 5 and the ratio of reference-to-signal beam intensities determines the amplitude of the input signal. Generally, a range of ratios from 2/1 to 10/1 is reasonable for good image reconstructions.

EFFECTS OF FILM NONLINEARITIES

In many applications, the recording exceeds the bounds of the linear range of the film transfer characteristics; consequently, the effects of the resulting non-linearity should be considered. This is done by using the model shown in Figure 6. It is assumed here that the waves considered vary only in one dimension. This assumption is made only to simplify the mathematical details that follow.

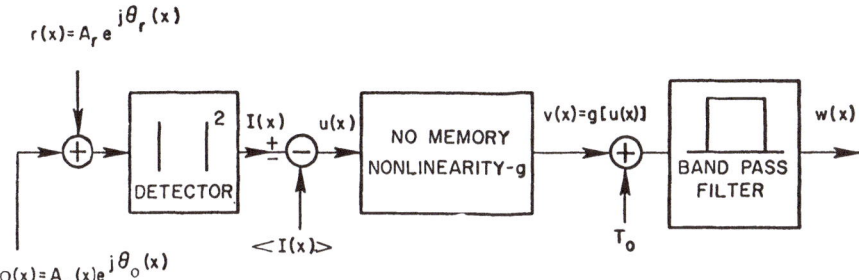

FIGURE 6. Model for representing nonlinear hologram recording

Let us consider a reference beam r impinging upon a recording media whose surface is in the x-y plane. Thus we have

$$r(x) = A_r \exp \left[j\theta_r(x) \right] \exp (j\omega_0 t) \qquad (10)$$

where $\theta_r(x)$ represents a spatial offset frequency times x and possibly a higher-order phase modulation of the reference beam. However, it is assumed that amplitude A_r is essentially constant over the surface of the recording media. Also, let the signal wave be represented by

$$o(x) = A_0(x) \exp \left[j\theta_0(x) \right] \exp(j\omega_0 t), \qquad (11)$$

where $A_0(x)$ and $\theta_0(x)$ represent, respectively, the spatial amplitude and phase modulation of the signal over the surface of the recording media.

Now the intensity of the wave formed by the sum of r(x) and o(x) is

$$I(x) = A_r{}^2 + A_0{}^2 + A_r A_0$$
$$\times \left\{ \exp[j(\theta_r - \theta_0)] + \exp [-j(\theta_r - \theta_0)] \right\},$$

which can be rewritten as

$$I(x) = A_r{}^2 + A_0{}^2 + 2A_r A_0 \cos (\theta_r - \theta_0), \qquad (12)$$

Here the time frequency ω_0 has been suppressed for convenience.

The quantity to be recorded is assumed to be proportional to the intensity I(x) times exposure time. The input to the nonlinearity u(x) is the intensity I(x) less the average value of I(x) (i.e., its bias level). The output of the nonlinearity is the functional $g(\cdot)$ plus some average level T_0.

Now an expression for the recorded image function v in terms of the applied waves r and s can be obtained by using the characteristic function method for a general odd nonlinearity g,

$$v(x) = \frac{1}{2\pi j} \int_{\sigma-j\infty}^{\sigma+j\infty} G(\xi)\Big(\exp\{A_r[2A_o \cos(\theta_r - \theta_o) + \alpha]\xi\}$$

$$- \exp\{-A_r[2A_o \cos(\theta_o - \theta_r) + \alpha]\xi\}\Big) d\xi \qquad (13)$$

for any $\sigma > \sigma_0$ when the unilateral transform

$$G(\xi) = \int_0^\infty g(x)\, e^{-\xi x}\, dx$$

exists for $Re(\xi) = \sigma > \sigma_0$ and $\alpha(x) = 1/A_r[A_o^2(x) - <A_o^2>]$. The exponentials can be expanded by using the Jacobi-Anger formula:

$$\exp(z \cos \theta) = \sum_{m=0}^{\infty} \varepsilon_m I_m(z) \cos m\theta, \qquad (14)$$

where ε_m is the Neuman factor $\varepsilon_0 = 1$, $\varepsilon_m = 2$ for $m = 1, 2, \ldots$, and $I_m(z)$ is a modified Bessel function of the first kind. Thus, Eq. (13) becomes

$$v(x) = \sum_{m=0}^{\infty} \sum_{n=0}^{\infty} \varepsilon_m \varepsilon_n \cos[m(\theta_r - \theta_o)]$$

$$\times \frac{1}{2\pi j} \int_{\sigma-j}^{\sigma+j} G(\xi)[I_m(2A_r A_o \xi)\, I_n(A_r \alpha \xi) \qquad (15)$$

$$- I_m(-2A_r A_o \xi)\, I_n(-A_r \alpha \xi)]\, d\xi$$

So that some specific results can be obtained, a full wave (odd) νth law device with $\nu \geqq 0$ is now assumed. Thus $g(x) = x|x|^{\nu-1}$, for $\nu \geq 0$. For the present, let us consider a signal that is composed of only two point sources. Hence,

$$o(x) = A_o(x) \exp[j\theta_o(x)] =$$

$$A_1 \exp[j\theta_1(x)] + A_2 \exp[j\theta_2(x)], \qquad (16)$$

where A_1 and A_2 are constant amplitudes so that $A_1 \gg A_2$. Such a signal occurs, for example, when one makes a hologram of two reflectors that differ greatly in reflectivity. Finally, it is assumed that only those terms of $v(x)$ for which $m = 1$ are of interest (that is, the other terms are assumed to be eliminated by spatial filtering). The above assumptions allow us to solve the integral indicated in Eq. (15)*. Thus, we finally have that

$$w(x) = K\{A_1 \cos(\theta_r - \theta_1) + [(1 + \nu/2]\, A_2 \cos(\theta_r - \theta_2)$$

$$+ \sum_{m=1}^{\infty} P_m(A_2/A_1)^{m-1} A_2 \cos[\theta_r - (m+1)\theta_1 + m\theta_2]$$

$$+ \sum_{n=1}^{\infty} q_n(A_2/A_1)^n A_2 \cos[\theta_r - (n+1)\theta_2 + n\theta_1] + \ldots\}, \qquad (17)$$

---

* For example see J. J. Jones, IEEE Trans., IT-9, 34 (1963).

where

$$K = \frac{4(A_r)^{\nu} \Gamma(\nu + 1)}{(1 + \nu)\Gamma^2 [(\nu + 1)/2]} A_1^{\nu - 1}$$

and p_m and q_n are functions of ν with the first few values given by $p_1 = -(1 - \nu)/2$,

$$p_2 = \frac{3 - 4\nu + \nu^2}{8} - \frac{1 - \nu^2}{8} \left(\frac{A_1}{A_r}\right)^2$$

and

$$q_1 = -\frac{1 - \nu^2}{8} \left[1 + \left(\frac{A_1}{A_r}\right)^2\right].$$

Thus if the signal is composed of two point sources, the output image will be comprised of an array of points. Two of these points are replicas of the original signal, and these are represented by the first two terms of the equation for w(x). The others, indicated by the two infinite sums, are simply false images. The first false image occurs at the location of the weak source's mirror image, which is at the same distance from the strong source as the weak source but on the opposite side. For a severe nonlinearity, this false image can have an amplitude as large as that of the reconstructed weak source. The other false images occur in pairs about the image of the strong source, so that each pair is more and more separated from the strong image. For a given nonlinearity, the magnitude of each pair of false images is proportional to progressively higher powers of the ratio of the weak-to-strong input signal. Thus, when the reflectivities of the two points differ greatly, only the first false image (or possibly the first three) need be considered. Furthermore, the relative amplitudes of the reconstructed signals depend upon the degree of nonlinearity of the film. In particular, for the case of an ideal limiter ($\nu = 0$) the ratio of strong-to-weak signal is increased by 6 db.

To demonstrate how film nonlinearities generate false images and suppress weak signals, several experiments were performed in which several holograms of two point sources were constructed. The input power ratio between the two points ranged from 10 db to 20 db, and the recordings were representative of both linear and nonlinear types.

FIGURE 7. Reconstruction of hologram of two-point target, linear recording.

For linear recording, the combined average intensity of the reference beam and the signal beam is located in the center of the linear portion of the curve representing the film's transmission vs exposure characteristics. Then, the variation

37

resulting from the modulating term is made sufficiently low that it is confined to the linear region. The nonlinear recordings are achieved by purposely modulating the intensity of the signal over a nonlinear portion of the films' characteristics.

FIGURE 8. Reconstruction of hologram of two-point target, soft nonlinear recording.

Some representative samples of holographic reconstructions of two point sources are shown in Figures 7, 8, and 9. As Figure 7 shows, only two points are observed in the case of the linear recording; the results for nonlinear recording, shown in Figure 8, depict an array of points exactly as predicted by the model shown previously. The two points located in the center of the array are replicas of the original signal and the other is simply a false image. It should be emphasized that this false image is not a higher order diffracted term because these are removed by the bandpass filter. When stronger nonlinearities were introduced, the number of observable false images was increased as shown in Figure 9. On comparing the intensity ratio of the strong signal to the weak signal in the reconstruction to that of the original signals, we noted the expected suppression of weak signals which measurement showed to range from 0 db to 2 db. In no case could we detect the hard limiter theoretical weak signal suppression of 6 db. This, of course, is attributable to the fact that none of the nonlinearities in our experiments were as severe as that of a hard limiter.

There is another deleterious effect of nonlinear recording that results in the deterioration of the reconstructed image. This is spectral folding, which involves the overlap of several harmonic orders. The result is a noisy or defocused image, which is observed with low offset frequency. We attempted to show the effect of spectral folding with an experiment in which we used a low offset frequency in constructing several holograms of a relatively large bandwidth signal. The nonlinearities of the film recording were progressively increased and the resulting reconstructed image in each case was investigated. Representative samples of the results are shown in Figures 10 and 11. Figure 10 shows the reconstruction of transparent targets on a black background. The recording in this instance was linear and the noisy background is the result partly of overlap from the signal spectrum located about zero frequency and partly of overlap from the conjugate image.

When the nonlinearities from recording increase, the background noise also increases, as is shown in Figure 11. This is caused by the additional overlap of nonfilterable, higher order harmonics, generated by the nonlinearities on the signal spectrum.

FIGURE 9. Reconstruction of hologram two-point target, strong nonlinear recording.

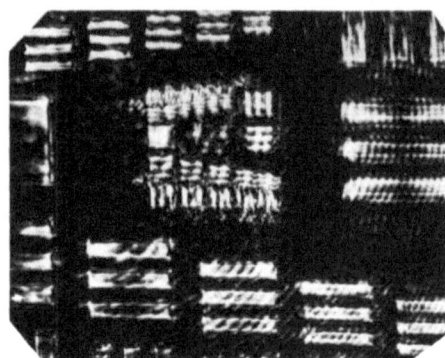

FIGURE 10. Reconstruction of hologram made with low offset frequency, linear recording.

FIGURE 11. Reconstruction of hologram made with low offset frequency, nonlinear recording.

EXPERIMENTAL CONSIDERATIONS

Holography involves sophisticated interferometric recording and, therefore, the stringent requirements of interferometry are applicable. Generally, this implies that illuminating sources have sufficient coherence, that the entire system be stable, and that the recording medium have sufficiently high resolution to enable recording of fine interference patterns. We shall now briefly consider the requirements.

6.1. Coherence Requirements

Stated simply, the illuminating source must have sufficient temporal and spatial coherence to enable recording of the fringes resulting from reference and object waves interference over a reasonable recording aperture. Temporal coherence limits the size of the object that can be photographed holographically. This coherence is generally characterized by a coherence length, $\Delta L \leq \frac{\lambda^2}{\Delta\lambda}$, where $\Delta\lambda$ is the wavelength spread of the source. The useful coherence length of the He-Ne laser at nominal power is approximately 30 cm. Generally, this length is sufficient for constructing holograms where the object dimension is relatively large. With the argon laser, however, the useful coherence for both the blue and green colors range from 6-20 cm at maximum and minimum mode full width respectively, thereby restricting somewhat the object size.

Recently, some techniques were introduced for increasing the coherence length. One such technique is the introduction of a Fabry-Perot etalon into the cavity. The coherence length is thus increased to some meters with a corresponding decrease in illumination intensity.

Spatial coherence denotes the ability of different parts of an illuminating beam to interfere. In general, the spatial coherence of gas lasers is sufficient for holography since most such lasers operate in the TEMoo mode. Those operating with transverse multimodes have only limited spatial coherence. In pulse laser holography, the poor spatial coherence of the pulse laser presents a major obstacle.

6.2. Stability Requirements

Once the light emanating from a source is divided into two beams, one representing the reference beam and the other the signal beam, the difference in optical path length must remain stationary for the duration of the exposure. The optical path should not change by more than $^1/_4$ wavelength of the illuminating light. With reasonable precautions, this severe requirement may be readily met. These may be itemized as follows:

1) Place optical system on a floating platform (rubber inner tubes are usually sufficient) to isolate it from ground and building vibrations.
2) Use relatively massive components and avoid the use of springs.
3) Place an enclosure over the entire system to maintain reasonable temperature and pressure stability.

6.3. Film Resolution Requirements

The recording medium must have sufficient resolution to record the finest interference fringes. This may be calculated from a knowledge of the maximum angle subtended between the reference beam and signal beam. If we refer to this angle as θ_{max} then the required film resolution is

$$d_{film} \geq \frac{\lambda}{2 \sin\left(\frac{\theta_{max}}{2}\right)} , \qquad (18)$$

40

For three-dimensional objects, the required resolution is in range of 0.5μ to 1.0μ.

THREE DIMENSIONAL RECORDING MEDIA

Holograms recorded in three dimensional media have distinctive properties when compared with conventional planar holograms:

1) The inherent high angular and wavelength sensitivies provide high data storage capabilities, and

2) Appropriate choice of thickness will allow high diffraction efficiencies. The interference effects are recorded as surfaces within the recording medium, so as to form in effect a three dimensional grating. The diffraction process from such a structure is analogous to x-ray diffraction from crystals, and has in fact been studied in this context.

The wavelength or the direction of incidence of the readout beam may be varied with a corresponding change in diffraction intensity from a hologram. Both of these parameters can therefore be effectively used to construct a hologram so as to store a great multiplicity of images, each separately stored uniformly throughout the recording medium. The more convenient technique is that of varying the direction of incidence of the illumination beam whilst maintaining the wavelength constant.

For the case of a simple grating hologram, which was constructed by recording the interference of two plane waves, some interesting relations were derived by adopting existing x-ray diffraction theory. These are given in terms of the diffraction efficiency of a hologram, which is defined as the ratio between the diffracted intensity and the incident intensity of the reconstructing beam. For thick absorption holograms,

$$\frac{I_d}{I_i} \approx e^{-\mu_0 \frac{t_0}{\cos \theta_B}} \left\{ \frac{\sinh^2 \left[\mu_1 \frac{t_0}{2\cos \theta_B} \sqrt{1 - \left(\frac{\pi\alpha}{\mu_1\lambda}\right)^2} \right]}{1 - \left(\frac{\pi\alpha}{\mu_1\lambda}\right)^2} \right\} \tag{19}$$

where

$$\alpha = 2(\theta_B - \theta) \sin 2\theta_B; \quad \lambda = \lambda_B$$

or

$$\alpha = 4 \sin^2 \theta_B \left(\frac{\lambda - \lambda_B}{\lambda_B}\right); \quad \theta = \theta_B$$

The average absorption coefficients μ_0 and μ_1 are associated with the incident and diffracted beams respectively, t_0 is the actual thickness of the hologram, θ_B is the Bragg angle, λ_B the Bragg wavelength, θ is the incident angle, and λ is the incident illumination wavelength. The maximum diffraction value occurs at $\theta = \theta_B$ and $\lambda = \lambda_B$ so that Eq. (19) becomes

$$\frac{I_d}{I_i} \approx e^{\mu_0 \frac{t_0}{2\cos\theta_B}} \quad \sinh^2\left(\mu_1 \frac{t_0}{2\cos\theta_B}\right) \tag{20}$$

The minimum loss possible for a given μ_1 is obtained when $\frac{\mu_0}{2} = \mu_1$, which when substituted into equation (20) leads to an optimum theoretical diffraction efficiency of $1/27$ or 3.7%.

The general expression (19) may also be used to describe the nature of the diffraction. After some algebraic manipulations, the general expression leads to a solution for the angular width between the half-power point of the curve of diffracted intensity vs. readout beam angle,

$$\Delta\theta_{1/2} = \sqrt{\frac{\ln 2}{\pi}} \frac{2\lambda \cos\theta_0}{t_0 \sin(\theta_0 + \theta_r)} \tag{21}$$

where the angles θ_r and θ_0 denote the internal incidence angles of the reference beam and signal beam each impinging from the two sides of the normal to the hologram.

The effect of the refractive index of the medium is derived by applying Snell's relation. The external half-power width, $\Delta\theta'_{/2}$, is then found in terms of the external parameters,

$$\Delta\theta_{1/2\,\text{ext.}} = \sqrt{\frac{\ln 2}{\pi}} \left(\frac{2\lambda'}{t_0}\right)$$

$$\left[\frac{\sqrt{n^2 - \sin^2\theta'_r}\,\sqrt{n^2 - \sin^2\theta'_0}}{\cos\theta'_r\left(\sin\theta'_0\,\sqrt{n^2 - \sin^2\theta'_r} + \sin\theta'_r\,\sqrt{n^2 - \sin^2\theta'_0}\right)}\right] \tag{22}$$

where all the primed parameters denote external parameters, and n is the index-of-refraction of the medium.

Some experiments have been performed with a variety of three dimensional recording medium. These included thick photographic emulsions and photochromic materials. The latter offer particular advantage in data storage because of their relatively high thickness and excellent resolution capability. For example, a 1 cm cubed photochromic crystal with resolution elements of $1/3000$ mm contains 27×10^{12} resolution cells.

An example of the angular orientation sensitivity of a photochromic material is shown in Figure 12. Here is shown the relative diffracted intensity as a function of incidence angle on the readout beam for a F centered KBr crystal. The experimental parameters in this case: crystal thickness was 4 mm, the reference beam and signal beam incident angles θ_r and θ_0 respectively were each $30°$. As shown the angular bandwidth between the half-power points is 1 min of arc. It is interesting to note that this value is significantly lower than $10°$ which is typical for conventional photographic emulsions.

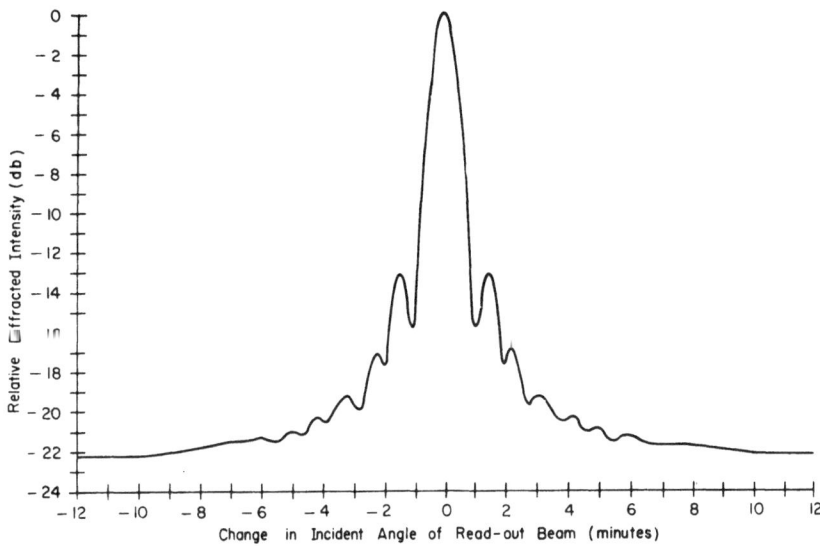

FIGURE 12. Experimental results of angular orientation sensitivity of photochromic recording materials. Thickness = 4 mm; Offset angle = 60°; $\theta_r = \theta_0 = 30°$.

MULTICOLOR HOLOGRAPHY

The optical arrangement for producing holograms which yield multicolor imagery is similar, albeit more sophisticated, to conventional hologram systems. The essential change is that instead of monochromatic illumination, coherent light of several wavelengths (generally red, green and blue) is used in producing the hologram. Each wavelength generates an independent fringe system that is recorded on the same photographic film or plate. Reconstructed images may be observed by illuminating the hologram with the same colors. Since most possible colors can be duplicated by combinations of the three primary colors, the viewer sees a striking reconstruction of the photographed scene possessing three dimensions, parallax, and exact color balance.

The reconstruction process has some complications. For the tricolor case we essentially have three holograms recorded on black and white film. In the reconstruction, each beam will interact with the interference patterns which are not only produced by its own wavelength, but by the other wavelengths as well. Thus, a number of spurious and generally overlapping images are generated in addition to the true multicolor image. The prevention or at least attenuation of these additional images, commonly referred to as ghost images, is the major problem of multicolor holography. We devote the major part of this section to the discussion of the ghost images and the various methods used for their elimination.

8.1. Analysis

In order to understand the details of the reconstruction process, we proceed by considering a polychromatic wavefront scattered by an object, given by

$$o = \sum_{i=1}^{n} o_i$$

and a polychromatic reference wavefront

$$r = \sum_{j=1}^{n} r_j$$

Here each of the o_i and r_j is assumed to represent complex wave amplitudes, corresponding to wavelengths λ_i, λ_j. These wavelengths, in turn, belong to the set $(\lambda_1, \lambda_2, \ldots, \lambda_n)$ of n available wavelengths.

After making the usual assumption that there is no coherence between the various spectral components, we let the wavefronts interfere at the hologram plane and derive the intensity distribution I. The resultant operation is mathematically described as

$$I = \left| o + r \right|^2 \delta_{ij} = \left| \sum_{i=1}^{n} o_i + \sum_{j=1}^{n} r_j \right|^2 \delta_{ij} \tag{23}$$

where δ_{ij} is the Kronecker delta function. After rearrangement, Eq. (23) becomes

$$I = \sum_{i=1}^{n} \left(\left| o_i \right|^2 + \left| r_i \right|^2 + o_i^* r_i + o_i r_i^* \right). \tag{24}$$

The intensity is recorded on a photographic plate which is developed so that the amplitude transmission of the plate at any point is proportional to I. Thus, when the hologram is illuminated with a polychromatic beam of the same kind as before,

$$c = \sum_{j=1}^{n} c_j,$$

a wavefront H emerges from the hologram. This wavefront is given by

$$H = \sum_{j=1}^{n} c_j \sum_{i=1}^{n} \left(\left| o_i \right|^2 + \left| r_i \right|^2 + o_i^* r_i + o_i r_i^* \right) \tag{25}$$

The first two terms inside the parentheses give rise to the zero-order terms during reconstruction and are of no interest. The last two terms represent the two first-order images;

$$\sum_{i=1}^{n} \sum_{j=1}^{n} o_i^* r_i c_i$$

44

represents the real image and

$$\sum_{i=1}^{n}\sum_{j=1}^{n} o_i r_i^* c_i$$

the virtual image. Now we consider the process in greater detail. To simplify the analysis, we assume that all the wavefronts originate from point sources and that their intensities are constant across the hologram. Therefore we neglect the amplitude and deal with only the phases of all the participating wavefronts.

In the following equations, x_0, y_0, z_0 are the coordinates of a multicolor object point; x_i, y_i, z_i are the coordinates of the reference point sources; and x_j, y_j, z_j are the coordinates of reconstructing illuminating sources. All these sources generate spherical wavefronts, and the subscripts i and j are associated with a source of particular color component in the recording and in reconstruction of the hologram, respectively. The chosen coordinate system has its origin in the center of the hologram with the x and y axes in the hologram plane. The positions of the subject points are illustrated in Figure 13.

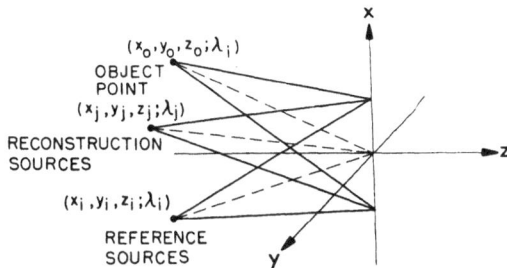

FIGURE 13. Illustration of object and source locations relative to a hologram plate located in the xy plane about the origin.

Some useful relations are readily derived from first-order analysis. For the case of multicolor wavefront reconstruction where n wavelength components are used in recording and n components in reconstruction, there will be $2n^2$ Gaussian image points. Half of these represent what is normally referred to as the virtual image V, and the other half the real image R.

By extending monochromatic analysis to include the use of multiple wavelengths, the phases ψ_{ij} of the reconstructed wavefronts at the exits of the hologram in the reconstruction process are

$$\phi_{ij} = \frac{2\pi}{\lambda_j}\frac{1}{2}\left\{(x^2 + y^2)(1/z_j) \pm (u_{ij}/z_o) \mp (u_{ij}/z_o)\right.$$

$$- 2x[(x_j/z_i) \pm (u_{ij}x_o/z_o) \mp (u_{ij}x_i/z_i)]$$

$$\left. - 2y[(y_j/z_i) \pm (u_{ij}y_o/z_o) \mp (u_{ij}y_i/z_i)]\right\}, \qquad (26)$$

where the upper sign refers to the virtual image and the lower sign to the real image; the wavelength ratio of reconstructing to recording light is $u_{ij} = \lambda_j/\lambda_i$.

Similarly, by extending the monochromatic analysis to include polychromatic wavelengths, the corresponding equations for the angular, lateral, and longitudinal magnifications for each of the reconstructed image points are given by

$$M_{ang\ ij} = \pm u_{ij},$$

$$M_{lat\ ij} = [1 \pm (z_o/u_{ij}z_j) - (z_o/z_i)]^{-1} \qquad (27)$$

$$M_{long\ ij} = -(1/u_{ij})\left[\left\{1 - z_o[(1/u_{ij}z_j) + (1/z_i)]\right\}^2\right]^{-1}$$

$$= -(1/u_{ij})M^2_{lat\ ij}.$$

To illustrate the importance of these parameters, let us consider a specific example. Assume that equal tricolor illumination is used for both recording and reconstructing the hologram, that is $\lambda_i = \lambda_j = (\lambda_1, \lambda_2, \lambda_3)$. These wavelengths may correspond to the three primary colors, red, green, and blue. Let us also suppose that the reference beams and reconstructing beams for all three wavelengths originate from the same point. Then $(x_i, y_i, z_i) = (x_j, y_j, z_j) = (x_1, y_1, z_1)$.

Using these assumptions and Eqs. (26) and (27), we determine that nine virtual images will be observed. Three of these images, obtained by putting $i = j$, correspond to points of the three colors superimposing exactly upon one another and forming a reconstructed point image that is an accurate image of the multicolor object point source. The corresponding magnifications for these three points are also the same; that is,

$$M_{ang\ 11} = M_{ang\ 22} = M_{ang\ 33},$$

$$M_{lat\ 11} = M_{lat\ 22} = M_{lat\ 33}, \qquad (28)$$

$$M_{long\ 11} = M_{long\ 22} = M_{long\ 33}$$

The other six images, obtained by putting $i \neq j$, represent reconstructed image points, hereafter referred to as ghost images, each of a single color. All ghost images are laterally and longitudinally displaced from the true image point and magnified according to their respective wavelength ratios u_{ij}. These ghost images are undesirable because they tend to distort the color reproduction process.

8.2.   The Different Methods of Multicolor Holography

In this section we briefly survey some of the methods which have been proposed and demonstrated in multicolor holography. The underlying result that each method is attempting to achieve is the elimination of the ghost images. For a detailed description of these systems, the reader should consult the references.

8.2.1. Frequency Multiplexing. One of the earliest proposed methods utilizes spatial frequency multiplexing. Each color component of the recorded signal is modulated onto a different spatial carrier so that the signal for each color occupies a different spatial frequency band. This is achieved by letting the orientation of the reference beams of each color impinge on the hologram from a different orientation.

In the reconstruction, the hologram must be reilluminated by reference beams of the same wavelength and orientations with respect to the hologram plate as during the construction. With proper geometrical arrangement, the ghost images are sufficiently displaced from the "true" multicolor image as to cause no problem. Optimum positioning, however, is a tedious task.

8.2.2. Spatial-Multiplexing. Another method to eliminate the ghost images is to form a hologram in such a way that the different interference patterns of each wavelength are not allowed to overlap on the emulsion. That is, we spatially multiplex, in a nonoverlapping manner, the several holograms corresponding to the various colors used. This may be achieved, for example, by overlaying a mask consisting of many small red, green, and blue color filters on the hologram during exposure. In reconstruction, the same mask-hologram arrangement is used so that any given color illuminates only that portion of the hologram which had been formed with that color.

8.2.3. Coded Reference Beam. Another effective technique for reducing the effects of the ghost images involves the coding of the reference beam. Both the amplitude and phase of the reference wave impinging on the hologram plate have decidedly different characteristics for each of the colors used to make the hologram. This can be simply achieved by placing a diffused glass in the path of the reference beam, thereby coding the reference wave. When reconstructing the multicolor image, the developed hologram must be accurately replaced in the position it occupied during recording, and reilluminated with the identical coded reference beam used in its formation. For proper choice of coding medium, this process is highly wavelength sensitive, thereby eliminating the ghost images.

8.2.4. Limited Field of View. For objects where the field of view angle is relatively small, acceptable results may be obtained without resorting to any special precautions. For reasonably separated illumination wavelengths, the ghost images fall outside the region of the desired multicolor image. A collinear arrangement in which all three wavelengths are contained in a single beam is utilized, thus facilitating the recording and reconstructing processes.

8.2.5. Thick Recording Media. One of the more convenient methods applicable to large field objects utilizes Bragg effects to suppress the ghost images. This stems from the fact that a hologram made on thick emulsion behaves as three-dimensional gratings which possess both angular and wavelength sensitivities.

The wavelength sensitivity of thick holograms may be represented in terms of the diffracted intensity at the half power points. This wavelength bandwidth, derived from three-dimensional grating theory is given in terms of

$$\Delta\lambda_{1/2} = \sqrt{\frac{\ln 2}{\pi}} \ \frac{\lambda^2 \cos \theta_0}{t_0 \sin^2\left(\frac{\theta_0 + \theta_r}{2}\right)} \tag{29}$$

where λ is the illuminating wavelength,

 t_0 is the thickness of recording medium,

 θ_0 is the average angle subtended between object and photographic plate normal, and

 θ_r is the angle subtended between reference beam and plate normal.

For typical angles and 15μ thick emulsion, this wavelength bandwidth is greater than 1000Å. For back reflection hologram in which the reference beam and object beam impinge on the hologram plate from opposite sides, the bandwidth is significantly reduced to approximately 100Å.

For thicker materials, such as photochrmics, with thickness ranging to 8000 microns, we get corresponding increase in wavelength sensitivity; wavelength bandwidth values of 10Å or so have been easily achieved with transmission holograms.

8.3. *Experimental Procedure and Discussion*

In our experiments we concentrated on the technique which utilizes thick recording materials to attenuate the ghost images. For this case, the required experimental setup for both construction and reconstruction of the holograms is shown in Figure 14. As shown, it is a relatively simple and convenient arrangement.

Two spectral components (4880Å and 5145Å), simultaneously generated by a Raytheon Argon Laser Model LG-12, were combined with the light (6328Å) from a Spectra-Physics Helium Neon Laser Model 125 by a b am splitter. The optical configuration was arranged to produce two beams from the beam splitter, each containing three spectral components. One beam was used as a reference beam and the other to illuminate the three-dimensional object. The scattered waves from the object interfered with the reference waves, and the resulting three interference patterns were recorded on Kodak 649F spectroscopic plates having an emulsion thickness of about 15μ after development.

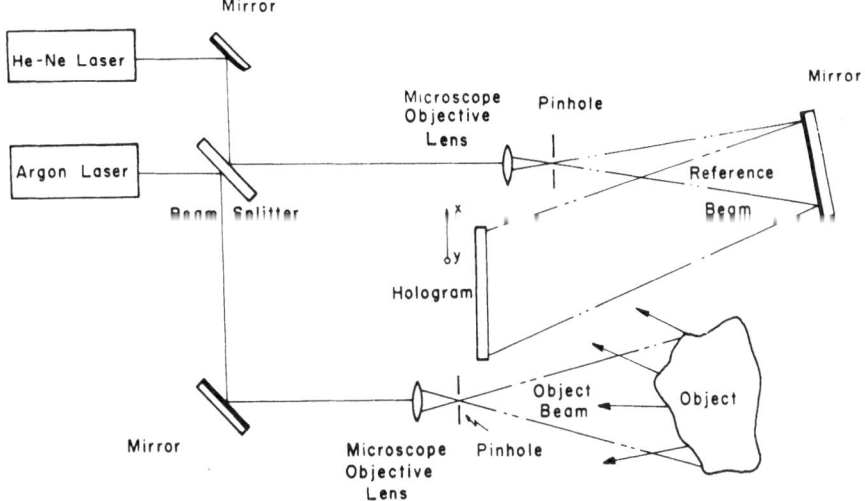

FIGURE 14. Multicolor wavefront reconstruction-hologram recording system.

Both sequential and simultaneous recording of the independent wavelengths were performed. The results were equivalent but the sequential method is more convenient. It allows independent object- and reference-beam control for each illuminating wavelength. This is necessary for proper recording of each of the three independent fringe systems in order to achieve the correct color balance in the reconstruction.

In addition to maintaining an accurate color balance in object and image by adjusting the intensity of each wavelength during recording and reconstruction, two precautions were observed throughout the experiments. First, the angle between the reference and object beams was generally above 100° and second, the hologram plate was oriented so that the normal to the plate bisected this angle. The former precaution was designed to decrease the spacing between the recorded fringes (less than 0.5μ), thereby increasing the selectivity of the hologram for different wavelengths and reducing the cross-talk between the various colors. The latter precaution produces recorded fringe surfaces that are, on the average, perpendicular to the emulsion surface, thus minimizing the deleterious effect of emulsion shrinkage which occurs during development.

In the reconstruction, the hologram was illuminated by a replica of the reference beam and was oriented to satisfy the Bragg angle for the three wavelengths simultaneously. The result is an impressive multicolor reconstruction that possesses the three-dimensional characteristics of the original object.

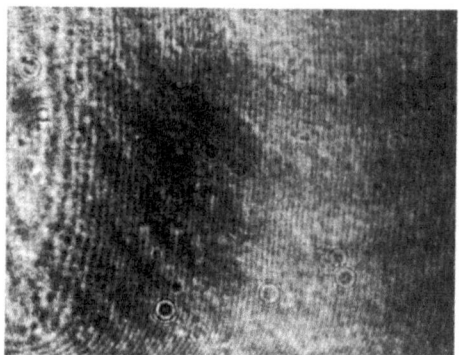

FIGURE 15. Typical appearance of a hologram.

The effectiveness of the process is illustrated by Figure 15 and Plate I. Figure 15 is a photograph of a typical hologram. Although this hologram, when reconstructed, will exhibit a multicolor image, its physical appearance cannot be distinguished from that of a conventional single-color hologram. All holograms used in these experiments were recorded on Kodak spectroscopic 649F plates with emulsion thicknesses of 15 microns

Plate I, A* is a photograph of a reconstruction of a toy car, 15 cm wide, 10 cm high, and 8 cm deep. Only red and blue illuminating colors, 6328Å and 4880Å, were used in recording and reconstructing. The separation in wavelengths, 1448Å, is sufficiently great so that no ghost images can be observed in the reconstruction.

Plate I, B shows an example of a tricolor reconstruction. The dimensions of the three-dimensional hexagonal vase were 18 cm in height and 13 cm in width and depth. The published print was made by photographing the virtual image with a conventional camera with aperture set at f/8 in order to provide an adequate depth of field to focus the entire scene. As can be observed in Plate I, B ghost images, resulting from the cross-talk between the 4880Å and 5145Å wavelengths of the Argon laser, were not completely eliminated. The closeness of these wavelengths in conjunction with insufficient emulsion thickness of 649F plates causes these deleterious images. Upon closer scrutiny it was noted that the size of each ghost image differed from the multicolor image size. These magnifications, as well as the ghost image displacements were predicted in the analysis of Section 8.1. It was also found that, although the color rendition of the reconstructed image duplicates that of the original object when illuminated with a tricolor laser beam, it is not the same as if the object were viewed by ordinary white light. For example, we could not

---

* Plate I is located between pp. 56-57.

reproduce most yellows. This, however, can be predicted from colorimetry charts for the three wavelengths used.

It is interesting to note that visual observation of the reconstructed images indicated that the specular characteristic normally associated with single-color hologram reconstructions is reduced in the multicolor reconstructions. We believe that this is because all three wavelengths are mutually noncoherent. Thus, specular effects tend to average out in the superposition of the three single-color reconstructed images.

APPLICATIONS

Numerous applications have been proposed for holographic photography. Some of these are currently being developed at many laboratories. In this section, a brief description of some of these applications is given.

9.1. Microscopy

At visible wavelengths, magnification without the use of lenses is achieved by diverging the light beams in construction of the hologram or during reconstruction. As shown with Eq. (8), the reconstructed image can be greatly magnified, and is capable of focusing an image at any plane in the depth dimension. An illustration of the magnification capability of holography is shown in Figure 16. Here we see a portion of a test pattern which is magnified by 120 using holographic methods. The estimated resolution is 3 microns. Although lenses are not used, the hologram (which behaves as a lens) gives rise to the aberrations and distortion found in conventional microscopy. This is quite apparent at higher magnifications.

As lasers are extended to the non-visible wavelengths, hopefully to the x-ray region, holographic microscopy offers considerable advantage. Enormous magnification is possible, produced in part by diverging beams and in part by the wavelength ratio between x-rays used in producing the hologram and the visible light used in the reconstruction.

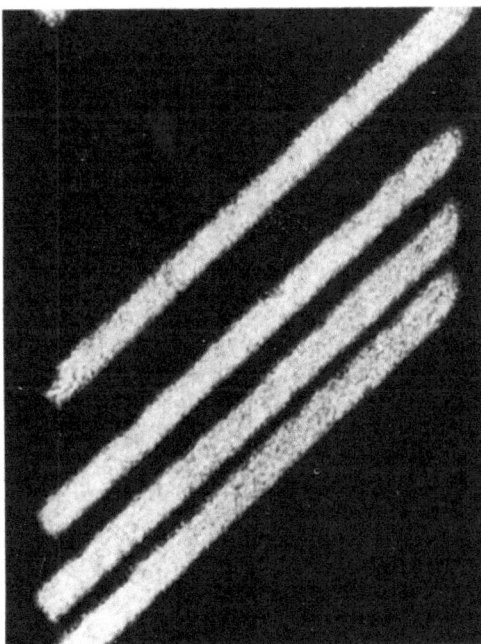

FIGURE 16. Holographic application in microscopy-lenseless magnification × 120 of a test pattern (courtesy of Leith and Upatnieks).

9.2. *Visual Displays*

Holography is potentially a powerful scheme in visual displays and simulation. In fact, it has already been commercially introduced in a number of advertising displays. At the University of Michigan we made several holograms as large as a foot square. Many problems must be overcome before visual displays can be widely used, however. One such problem is finding projection techniques to allow observation of reconstructed images by large audiences.

The reconstructed real image is sometimes more desirable in display applications. In a sense it is more striking than the virtual image, since the observer can "approach" each element of the scene as closely as he wishes. Unfortunately, the real image is pseudoscopic and is fraught with many conflicting visual cues: near objects are obscured by more distant ones, concave surfaces appear convex, etc. Consequently, it is difficult to view the real image. A technique of constructing two holograms is succession overcomes this difficulty. The pseudoscopic real image of the first hologram is re-inverted with the second hologram. In the resulting real aerial image, all the aforementioned visual anomalies are absent.

9.3. *Hologram Interferometry*

Holographic interferometry is one of the most important applications for holography. The underlying principle is that the reconstructed image duplicates the original object in amplitude and phase. The duplication is so precise that reconstructed and actual image fuse together and appear as one. The reconstructed image may be substituted for the actual object in an interferometric application, say use the reconstructed image as a "standard" to test other similar objects. Similarly, the reconstructed image and actual object can be interfered against each other thus allowing sensitive detection in optical path whether distance or index of refraction.

Several proposed and demonstrated applications include the measurement of vibrations, stress, and strain and the detection of flow patterns in a wind tunnel or a shock tube. Figures 17 and 18 illustrate this application. The photographs in Figure 17 show reconstructions from holograms of a vibrating can bottom. The image has superimposed upon it contours which can be interpreted as loci of equal amplitude of vibration. Figure 18 shows a photograph of a reconstructed image from a doubly exposed hologram. The first exposure was of the soldering iron at room temperature and the second after it was heated for five minutes. From the fringe formation quantitative data about the changes in index-of-refraction of the surrounding air may thus be obtained.

An interesting extension of the interferometric techniques results from the use of several object-illumination source positions or several different frequencies in the hologram construction process. This produces a reconstructed image with superimposed contours of constant depth.

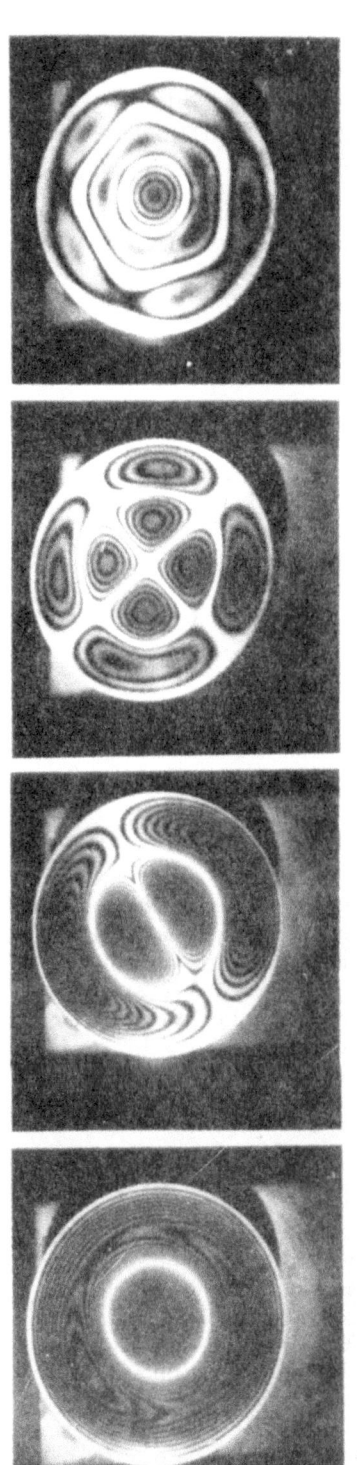

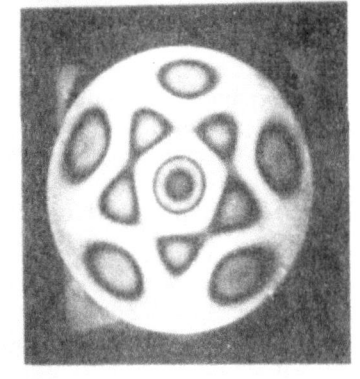

FIGURE 17. Holographic vibrational contours of six different modes of vibrating can bottom (courtesy of Powell and Stetson).

53

For the case of two frequencies, if the illumination source is placed anywhere along the line of sight, then the distance between contours is

$$d = \frac{1}{2} \frac{\lambda_1 \lambda_2}{\lambda_1 - \lambda_2}$$

(30)

Using this technique, depth contours with fringe spacing of 10 microns can be readily attained in the laboratory.

FIGURE 18. Holographic interferogram of natural convection in air.

9.4. Computer-Generated Holograms

Many laboratories are actively engaged in investigations of computer-generated holograms. The motivation is to generate images of objects which do not exist physically but are known in mathematical terms. A computer is used to calculate the diffraction pattern. The resulting information is fed into an automatic plotter making a large scale drawing which is then photographically reduced resulting in a hologram. As demonstrated, these synthetic holograms behave like conventional holograms, although their quality thus far is significantly poorer.

9.5. Data Reduction of Holographic Information

The large bandwidth required for transmission of holographic data is impractical for existing systems. A recently proposed technique alleviates this problem to a large extent. The light from the object is diffracted by an intermediate dispersion plane such as a diffuser and collected at the hologram aperture. Correct illumination of the hologram provides an image beam which passes back though the intermediate medium and comes to focus in the space originally occupied by the object. The hologram aperture may be made extremely small thus representing a decrease in the bandwidth required for transmission of the data. As with earlier techniques, the image suffers from degradation in resolution or S/N ratio.

Trade off between resolution and S/N ratio can be controlled with proper choice and location of diffusing medium. An important advantage of this technique over others is that the viewing angle in not affected. Space-bandwidth reduction of 3000 was already achieved with this technique.

9.6. Coding

Although the hologram is a kind of coded message, the information content can be easily retrieved with a coherent light source. The information recorded could be made completely secure, however, by placing between the object and the hologram a diffuser which may be ground glass or generally some translucent material. From such a hologram we cannot reconstruct the original object unless we have an identical diffuser and have a "a priori" knowledge of the hologram diffuser positional relationship when the hologram was made.

9.7. Incoherent Illumination Holography

Construction of holograms with incoherent radiation offers exciting possibilities. For the case of visible light, it will allow the recording of holograms by the layman; but the more promising possibilities are in radiation where coherent sources are not available — say x-rays.

Several techniques have been proposed for the construction of holograms with incoherent light. In all the proposed methods, the incoherently illuminated image is split into two displaced images, which interfere with each other and recorded as a hologram. This operation is performed with a suitable interferometer so that the corresponding points in the two images which are coherent with each other produce interference fringes whereas other parts of the image will form a uniform bias. The basic problem which exists in incoherent holography is the rapid buildup of the ratio of the bias to the fluctuating intensity as the number of points in the object is increased. As this ratio is increased, the degradation of signal to noise in the reconstructed image becomes intolerable. To date, only objects of relatively small number of points such as transparencies of targets on a black background can be reconstructed.

9.8. Holographic Imaging Through Turbulent Media

Imaging through random media, say the atmosphere, has received considerable attention in the past. With conventional image forming systems, the phase perturbations of the media degrade the image to an intolerable degree. Under some conditions, a recent scheme proposed by a Stanford University team overcomes some of the problems. Let a hologram be formed in the vicinity of the turbulent media represented by $e^{i\phi}$. By passing the reference beam, U_o and object beam U, through the random medium, they both acquire the same phase modulation $e^{i\phi}$, so that the resulting intensity is

$$I = \left| U_o e^{i\phi} + U e^{i\phi} \right|^2 = \left| U_o + U \right|^2 \tag{31}$$

Thus, the hologram record is unaffected by the random media, and a reconstructed image free from defect is thereby reconstructed.

9.9. *Pulse Laser Holography*

The use of pulse lasers in holography offers high-power short-exposure capability enabling holographic recording of moving three-dimensional objects. It also enables the recording of transient phenomena which were hitherto impossible with conventional holography.

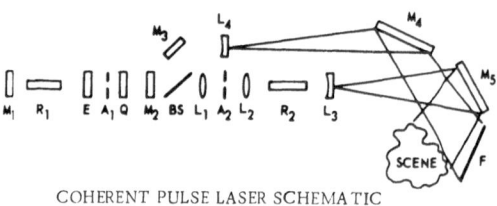

COHERENT PULSE LASER SCHEMATIC

FIGURE 19. A single-mode pulsed laser experimental setup for making holograms (courtesy of Siebert).

The principle difficulty in making pulse laser holograms arises from the limited coherence of most pulse lasers, and although many interesting and exciting experiments were performed, they were limited to transparent objects. More recently, holograms made of front-lighted, three-dimensional scenes were reported. A single mode pulsed ruby laser was employed in conjunction with an amplifier. A diagram of the experimental setup is shown in Figure 19. The single-mode laser consists of end mirrors M_1 and M_2, optically pumped ruby rod R_1, passive Q switch, etalon E for selecting a single longitudinal mode, and aperture A_1 for selecting a single transverse mode. The mode selection results in a significant reduction of intensity so that an amplifier is required. The latter consists of lenses L_1, L_2, and L_3, aperture A_2, and an optically pumped ruby rod R_2.

This scheme was used successfully in constructing holograms of a running squirrel case rotor and animated subjects.

9.10. *Ultrasonic Holography*

Ultrasonic holography provides two important advantages over conventional ultrasonic imaging. First, it allows the observation of objects in three dimensions and secondly it overcomes, with proper location of reference beam and object beam, any deleterious medium turbulence. Aside from the existing problems in conventional ultrasonic imaging such as better sources and detectors, ultrasonic holography has additional problems. One major problem results from the transformation from ultrasonic waves to optical waves. This is particularly evident when

PLATE I ———►

A. Photograph of three-dimensional multicolor reconstruction of a model car. Wavelengths used in recording and reconstruction were 6328Å and 4880Å.

B. Photograph of three-dimensional multicolor reconstruction of vase. Wavelengths used in recording and reconstruction were 6328Å, 5145Å, and 4880Å.

dealing with three-dimensional objects. As shown by Eq. (8), the longitudinal magnification is the square of the lateral magnification resulting in image distortions during reconstruction.

Several techniques for holographic ultrasonic detection have been reported. Four of these probably form the basis for all techniques which are currently employed. The first is the liquid surface deformation technique. Here two sound waves, one reference beam and the other signal beam, are directed to the surface of a liquid to form an interference pattern. This pattern behaves as a reflecting phase grating which can be illuminated with laser light for immediate reconstruction, or photographically recorded for future reconstruction.

A second method is the scanning technique. In this technique the sound waves from the object are scanned by means of a transducer which converts the sound waves to an electronic signal. This signal is then interfered with an electronically generated reference beam and the resulting interference pattern is then recorded to form a hologram.

The third technique is referred to as acoustic imagery by Bragg diffraction. Although this imaging technique is not holographic, it has important similarities. Here the sound waves from an object travel in a liquid medium which behaves as a three-dimensional recording medium. The sound image is converted to an optical image by Bragg diffraction.

The fourth technique utilizes as a detector an ultrasonic camera. Here the sound waves from object and reference beam impinge on a quartz crystal located at the front windows of a cathode ray tube. The resulting interference is then displayed on a TV screen and photographed, thereby getting a hologram.

Because ultrasonic holography has very important applications in medical diagnosis, investigation of metal defects, and detection of three-dimensional objects in oceanographic displays, the field is being actively pursued.

CONCLUSIONS

The purpose of these lectures was to present the basic principles of holography, its potential uses, and its limitations. Obviously, all the applications which have been described in upward of 500 papers could not be presented in this series. The emphasis, therefore, was on the more immediate applications. In the future, we may witness revolutionary applications of holography, three-dimensional television, cinematography, and x-ray. However, at present, the relatively sparse research in these areas does not warrant fuller discussion.

Much more research must still be done before the relatively infant field of holography fulfills its potential. Hopefully, the material presented in the lectures has shed light on some aspects of holography and provided stimulation for further investigation.

REFERENCES

Basic Holography

D. GABOR, Proc. R. Soc., **A 197**, 454 (1949); Proc. R. Soc., **B 64**, 449 (1951).
E. N. LEITH and J. UPATNIEKS, J. opt. Soc. Am., **52**, 1123 (1962); J. opt. Soc. Am., **53**, 1377 (1963); J. opt. Soc. Am., **54**, 1295 (1964).

Magnification and Aberrations

E. N. LEITH, J. UPATNIEKS and K. HAINES, J. opt. Soc. Am., **55**, 981 (1965).
R. W. MEIER, J. opt. Soc. Am., **55**, 987 (1965).

Film Transfer Function

R. F. VAN LIGTEN, J. opt. Soc. Am., **56**, 1 (1966).
A. A. FRIESEM, A. KOZMA and G. F. ADAMS, Appl. Opt., **5**, 851 (1967).

Effects of Film Nonlinearities

A. KOZMA, J. opt. Soc. Am., **56**, 428 (1966).
A. A. FRIESEM and J. S. ZELENKA, Appl. Opt., **10**, 1755 (1967)

Three-Dimensional Recording Medium

DENISYUK, Optics Spectroc., **15**, 279 (1963).
VAN HEERDEN, Appl. Opt., **2**, 393 (1963).
A. A. FRIESEM, Appl. Phys. Letters, **7**, 102 (1965).
E. N. LEITH, A. KOZMA J. UPATNIEKS, J. MARKS and N. G. MASSEY, Appl. Opt., **8**, 1303 (1966).
D. R. BOSOMWORTH and H. J. GERRITZEN, Appl. Opt., **1**, 95 (1968).

Multicolor Holography

E. N. LEITH and J. UPATNIEKS, J. opt. Soc. Am., **54**, 1295 (1964).
K. S. PENNINGTON and L. H. LIN, Appl. Phys. Letters, **7**, 56 (1965).
L. MANDEL, J. opt. Soc. Am., **55**, 1697 (1965).
L. M. LIN, K. S. PENNINGTON, G. W. STROKE, and A. E. LABEYRIE, Bell System tech. J., **45**, 659 (1966).
J. UPATNIEKS, J. MARKS and R. J. FEDOROWICZ, Appl. Phys. Letters, **8**, 286 (1966).
A. A. FRIESEM and R J. FEDOROWICZ, Appl. Opt., **6**, 529 (1967).
E. MAROM, J. opt. Soc. Am., **57**, 101 (1967).
R. J. COLLIER and K. S. PENNINGTON, Appl. Opt., **6**, 1091 (1967).

Microscopy

E. N. LEITH and J. UPATNIEKS, J. opt. Soc. Am., **55**, 569 (1965).

Visual Displays

J. V. PARKER, Information Display, May/June 1966.
F. B. ROTZ and A. A. FRIESEM, Appl. Phys. Letters, **3**, 146 (1966).

Hologram Interferometry

R. L. POWELL and K. A. STETSON, J. opt. Soc. Am., **55**, 1593 (1965).
K. A. HAINES and B. P. HILDEBRAND, Appl. Opt., **5**, 595 (1966).
L. O. HEFLINGER, R F. WUERKER and R. E. BROOKS, J. appl. Phys., **37**, 642 (1966).
B. P. HILDEBRAND and K. A. HAINES, J. opt. Soc. Am., **57**, 155 (1967).
J. R. VARNER and J. S. ZELENKA, J. opt. Soc. Am., **58**, 723A (1968).

Computer-Generated Holograms

J. P. WATERS, Appl. Phys. Letters, **9**, 405 (1967).
A. W. LOHMANN and D. P. PARIS, Appl. Opt., **6**, 1739 (1967).

Data Reduction of Holographic Information

K. A. HAINES and D. B. BRUMM, J. opt. Soc. Am., **57**, 1412A (1967).
C. B. BURCKHARDT, J. opt. Soc. Am., **57**, 1412A (1967).

Coding

H. KOGELNIK, Bell Syst. tech. J., **44**, 56 (1965).
E. N. LEITH and J. UPATNIEKS, J. opt. Soc. Am., **56**, 523 (1966).

Incoherent Holography

A. KOZMA and N. MASSEY, J. opt. Soc. Am., **56**, 537A (1966).
G. COCHRAN, J. opt. Soc. Am., **56**, 1513 (1966).

Imaging Through Turbulent Media

J. W. GOODMAN, W. H. HUNTLEY, D. W. JACKSON and M. LEHMANN, Appl. Phys. Letters, **12**, 311 (1966).

Pulse Laser Holography

R. E. BROOKS, L. O. HEFLINGER and R. F. WUERKER, IEEE J. Quantum Elect., QE-2, 275 (1966).
L. D. SIEBERT, Appl. Phys. Letters, **10**, 326 (1967).

Ultrasonic Holography

R. K. MUELLER and N. K. SHERIDON, Appl. Phys. Letters, **9**, 328 (1966).
K. PRESTON and J. L. KREUZER, Appl. Phys. Letters, **10**, 150 (1967).
A. F. METHERELL, H. M. A. EL-SUM, J. J. DREHER and L. LARMORE, J. Acoust. Soc. Am., **4**, 733 (1967).
E. MAROM, D. FRITZLER and R. K. MUELLER, Appl. Phys. Letters, **2**, 26 (1968).

General Holography

E. N. LEITH and J. UPATNIEKS, Scient. Am. J., June 1966.
R. J. COLLIER, IEEE Spectrum, **7**, 67 (1966).
G. W. STROKE, Introduction to Coherent Optics and Holography, Academic Press (1966).
J. B. DEVELIS and G. O. REYNOLDS, Theory and Applications of Holography, Addison-Wesley (1967).

3/A Quick Look at Light Scattering with Laser Sources

S. P. S. PORTO

When light passes through matter weak random scattered radiation will appear. Leonardo da Vinci suggested scattering by particles in the air as the explanation of the blueness of the sky; this idea was pursued by many people including Newton and Tyndall who tried to identify the particles responsible for the scattering. It was Maxwell however who, studying the careful measurements of Lord Rayleigh, showed that the molecules themselves are responsible for the scattering and the subsequent studies led to Rayleigh's scattering theory [1], which gave correct interpretation to all known properties of scattering, i. e., the frequency dependence, critical opalescence, index of refraction, etc. As usually happens, when we think we understand a phenomenon well, most physicists turned their attention to other problems.

In 1922, Brillouin [2] predicted that if monochromatic radiation was allowed to scatter from an optical medium, side bands should appear, as a result, in essence, of a Doppler shift due to the reflection of part of the light beam by a sound wave going through the medium. The frequency shift should be a function of the angle of observation and of the sound velocity in the medium. In 1923 Smekal [3] considered, in the Bohr theory approximation, the scattering of light by a system that has two quantized energy levels and predicted, in essence, the effect later discovered in 1928 by Raman [4] and known as the Raman effect. It is interesting to note that working independently Landsberg and Mandelstam [5] in Russia discovered the Raman effect in quartz a few months after Raman's work. (The Russians still refuse to identify the effect with Raman's name, preferring the denomination "combinational scattering".) In 1930 Gross [6] confirmed the predictions of Brillouin, showing that the Doppler shifted frequencies appeared at the right frequencies, for both liquids and solids.

The flurry of enthusiasm over the Raman effect was at once evidenced. By 1934 more than 500 papers had been published on the Raman effect alone and by the early forties this number was a few thousands.

In 1934, Placzek [7] wrote a lengthy and excellent review paper on the Raman effect which had the same effect as the Rayleigh theory had had on the early work on light scattering. Most of the problems seemed to be understood and Raman effect papers in the Physical Review became rarer and rarer until the point in which they disappeared. The Raman effect became more and more a tool for the structural physical-chemist with only a handful of physicists, notably the groups under Krishnan, Raman and Baghavantam in India, Welsh in Canada, Mathlieu in France and Stekhanov in Russia, devoting most of their effort to the study of the Raman effect.

Light scattering had, in essence, a rebirth with the appearance of the laser sources. Not only now had the physical-chemist a new, more powerful and cleaner source available to him but the physicist had ways now of testing Placzeck's, Lorentz's and Brillouin's theories, could look for directional effects in scattering processes, could look for inelastic scattering processes which have very low cross sections, etc. The enthusiasm is evident by the number of laboratories and scientists now engaged in light scattering with laser sources.

In this paper we are going to try to show, from very simple arguments, the origin of some of the most common scattering mechanisms which are being studied today, without trying to be thorough and complete, since this would require a large book.

If an electric field E is applied on a medium having polarizability α, a polarization (most commonly but not necessarily a dipole) P will be induced obeying the relation:

$$\vec{P} = \alpha \, \vec{E} \quad \text{or} \quad \begin{pmatrix} P_x \\ P_y \\ P_z \end{pmatrix} = \begin{pmatrix} \alpha_{xx} & \alpha_{xy} & \alpha_{xz} \\ \alpha_{yx} & \alpha_{yy} & \alpha_{yz} \\ \alpha_{zx} & \alpha_{zy} & \alpha_{zz} \end{pmatrix} \begin{pmatrix} E_x \\ E_y \\ E_z \end{pmatrix} \qquad (1)$$

where α is a tensor, which means that if the field is applied in a certain direction i, the polarization can be induced in a different direction j. In the study of light scattering E is the field associated with an electromagnetic radiation and it can without loss of generalization be expressed as $E = E_0 \cos \omega_L t$ where ω_L is the source, or laser, frequency.

Since the atomic or molecular dimension is of the order of angstroms and the laser frequency is of the order of thousands of angstroms the usual approximation that the electric field is slowly varying across the molecular dimensions (Born approximation) is valid but there is also the fact that the field is constant over many molecular dimensions and what we sample, when we apply light on matter is an average polarization, and an average polarizability

$$\langle \vec{P} \rangle = \langle \alpha \rangle \vec{E}, \qquad (2)$$

We know that each molecule has a polarizability tensor $\alpha^{(i)}$ associated with it so if we neglect interaction between our molecules:

$$\langle \vec{P} \rangle = \langle N \, \alpha^i \rangle \vec{E} = V \langle \delta \rangle \langle \alpha^i \rangle \vec{E} \qquad (3)$$

i.e., the average polarizability of the medium can be approximated by averaging the product of the volume, the density of the sample, and the polarizability of each molecule.

We know also from electromagnetic theory that a permanent polarization such as a dipole will not interact with radiation but an oscillating dipole will emit or absorb light so to look for a scattering phenomena we have to look for a oscillating polarization, an electric dipole for instance.

Rayleigh and Brillouin scattering

In Eq. (3) let us concentrate on the density fluctuations and assume that the polarizability tensor associated with each molecule is a diagonal (trace) tensor, i.e., only α_{xx}, α_{yy} and $\alpha_{zz} \neq 0$. We can write those fluctuations in density as

$$\delta\rho = \left(\frac{\partial\rho}{\partial\rho}\right)_s \delta\rho + \left(\frac{\partial\rho}{\partial s}\right)_p \delta s , \qquad (4)$$

i.e., the density will have fluctuation with pressure or with entropy. The pressure fluctuations of the density are those, for instance, in which an acoustical wave propagates through the material while the entropy or thermal fluctuations will not propagate. The propagating pressure fluctuations will scatter the incoming photon at a displaced frequency and give rise to the Brillouin scattering while entropy fluctuations will give rise to the Rayleigh scattering. Imagine that the a sound, or acoustical, wave, characterized by a frequency ω_B and sound velocity v_B, travels through a medium. When this sound wave scatters a photon one has to have conservation of energy and conservation of momentum:

$$E_L = E_S \pm E_B \text{ or } \omega_L = \omega_S \pm \omega_B \text{ and } \vec{k}_L = \vec{k}_S + \vec{k}_B, \qquad (5)$$

The conservation of momentum diagram is shown in Figure 1 for all scattering processes. If we can assume small dispersion, i.e., $n(\omega_L) \simeq n(\omega_S)$, from Figure 1 we obtain $|k_B| \cong 2|k_L| \sin\frac{\theta}{2}$ or

$$|k_B| = \frac{\omega_B}{v_B} = 2\frac{\omega_L n}{C}\sin\frac{\theta}{2} \text{ or } \omega_B = 2\frac{v_B n}{C}\omega_L \sin\frac{\theta}{2} , \qquad (6)$$

So the Brillouin frequency shift obeys a Bragg or grating diffraction law and its measurement at a given angle and a given excitation frequency will give you a measurement of the sound velocity in the medium. Since the sound velocities in condensed matter are of the order of a thousand meters/sec v_B is of the order of 3-10 KMc/sec or about .1 - .3 cm^{-1} for 90° scattering if a visible laser is used for excitation. Experimentally the Brillouin and Rayleigh spectra are studied by observing the scattering in a Febry-Perot [9] interferometer, [8] a high resolution spectrograph, or by photobeating electronic techniques [10]. Since the normal line width of an argon laser is of the order of .15 cm^{-1} and that of a He-Ne is of the order of .05 cm^{-1} a lot of care has to be used to mode select the laser so that its line-width is less than that of the Brillouin line (~700 Mc/S).

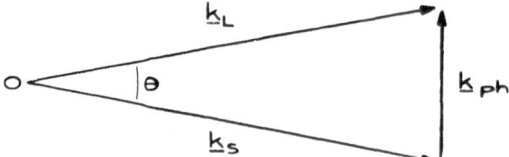

FIGURE 1. Conservation of momentum diagram in any scattering experiments. K_1, K_s and K_{ph} are respectively the wave vectors of the laser light, the scattered light and of the phonon or of any scattering quasi-particle.

In the way we arrived at the Brillouin and Rayleigh scattering as fluctuations in density, we can see that both Brillouin and Rayleigh lines are completely polarized. By relating the fluctuations in pressure and entropy to known thermodynamic quantities C_p and C_V one obtains the known Landau-Placzek relation between the intensities of the Brillouin to the Rayleigh scatterings:

$$\frac{I_B}{I_{Ray}} = \frac{C_V}{C_V - C_p} \quad . \tag{7}$$

For instance for water, where $C_V \cong C_p$, most of the intensity observed in the Rayleigh line, in low resolution instruments, corresponds to the two Brillouin components. Figure 2 shows a high resolution spectrum of the Rayleigh and Brillouin spectra of water.

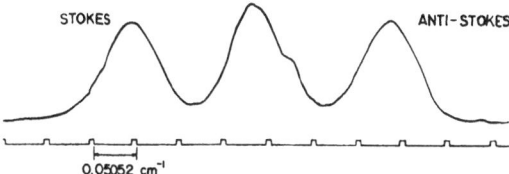

FIGURE 2. Brillouin scattering of water taken from [9] showing the two Brillouin components (outside lines) and the Rayleigh line of water. The Rayleigh line for water shown in the picture is almost all coming from scattering of particles suspended in the sample since for pure water the Rayleigh scattering should be very small.

Generalized Raman effects

For the purpose of this paper we are going to designate as Raman effects all the inelastic light scattering phenomena in which the scattering mechanism produces a change in the polarizability tensor associated with each molecule, as viewed in our laboratory frame of reference.

Rotational Raman effect

To each molecule we can associate a polarizability tensor which is tied to the symmetry axis of the molecule: x, y, z. Let us diagonalize this tensor and call the new tensor diagonals as α_1, α_2, α_3. In a completely spherical molecule $\alpha_1 = \alpha_2 = \alpha_3$ so if the molecule rotates, the tensor, viewed in the laboratory axis x', y', z' stays constant so that there will be no change, or modulation, of the polarizability and if there is no change in the polarizability there is no oscillating dipole so this rotation will be inactive in scattering.

If $\alpha_1 \neq \alpha_2$, for instance, the polarizability viewed in the laboratory system will change when the molecule rotates and the rotational Raman effect can be observed. The selection rules for the rotational Raman effect are $\Delta J = O$, ± 2 because each component of the tensor viewed in the laboratory system is equal to the sum of the components in the molecular system of reference multiplied by a factor containing two cosine functions.

The rotational Raman effect is completely depolarized (P=.75) and $\Delta J=0$ selection will give an undisplaced scattering (Rayleigh) which is polarized. The rotational frequencies are inversely proportional to the molecular moment of inertia and the rotational displacements are of the order of ~ 1 cm^{-1}.

To observe the rotational Raman effect high resolution spectrographs have been used almost universally up to date. Due however, to the fact that the Rayleigh scattering is so highly polarized while the rotational Raman effect is depolarized, one can foresee for the future the use of a laser source coupled with a single monochromator (a double monochromator will not be needed because in the right geometry the Rayleigh line is weak) and photoelectric techniques as the ideal way to observe rotational Raman effect [11].

Anisotropy Raman scattering in liquids

This kind of scattering is quite commonly known as the "Rayleigh wing" scattering which is observed in liquids. It is in a way very closely related to the rotational Raman effect and in a way also related to the Kerr effect in liquids. This anisotropy scattering is due to the fact that in a liquid the molecules sampled by the laser beam are rotating in a viscous medium and that one views a changing polarizability in the laboratory system of reference. This changing polarizability is due to rotation and also to changes of instantaneous aggregation states of the molecules. This "overdamped rotation" gives the same kind of polarizability changes responsible for the rotational Raman effect but there are now no discrete levels and the resulting spectrum is a low frequency continuum, around the laser exciting frequency.

Debye [12] has worked out details of this anisotropy scattering predicting a Lorentzian line shape for the scattering with a width that is dependent on the volume of the molecule, the temperature and the shear viscosity. For most liquids the half width of this anisotropy scattering is of the order of 5-10 cm^{-1}. Recent measurements of this anistropy scattering [13] show deviations from the predicted Lorentz shape.

This anistropy scattering even though hardly explored today should give much information on the angular correlation functions in liquids which are so badly needed to understand the nonlinear optical effects, like self-focusing, which are dependent on the Kerr effect.

Vibrational Raman effect in molecules

This is the oldest kind of Raman effect known and can be simply understood as a modulation of the polarizability tensor components due to a vibration of the molecule. Classically, if the polarizability is modulated, or if it changes with a vibration of the molecule

$$\alpha = \alpha_0 + \frac{\partial \alpha}{\partial q_m} q_m = \alpha_0 + \alpha_1 \sin \omega_M t \quad \text{so}$$
$$P = (\alpha_0 + \alpha_1 \cos \omega_M t)(E_0 \cos \omega_L t) = E_0 \alpha_0 \cos \omega_L t + \quad (8)$$
$$+ \alpha_1 E_0 [\cos(\omega_L + \omega_M)t + \cos(\omega_L - \omega_M)t]$$

We see from Eq. (8) that the polarization P will radiate energy at the frequencies $(\omega_L + \omega_M)$ and $(\omega_L - \omega_M)$, the anti-Stokes and Stokes-Raman vibrational frequencies, besides the Rayleigh scattering discussed before.

Group theory predicts the number of frequencies which are Raman, or infrared active, for all molecules if the shape of the molecule or its "point group" is known. Group theory will also predict for each normal mode which of the polarizability tensor components are changing during the vibrational motion measured by the Raman effect, and counting the number of frequencies which are Raman and infrared allowed, and measuring the depolarization of the Raman lines, one can gain considerable knowledge about the shape of the molecule under investigation. All molecular vibrations which are completely symmetric (such as the "breathing" motions) have changes in the diagonal components of the polarizability tensor and their scattering is polarized (depolarization ratio is close to zero) while all other normal modes will have changes mostly in the off diagonal terms of the polarizability tensor or have the trace of the tensor equal to zero and their scattering consists in a depolarized Raman effect (depolarization ratio = .75).

It is interesting to mention here the "vibrational overtone" Raman effect. We know that the selection rule for a vibrational Raman effect is $\Delta v = \pm 1$; an overtone Raman effect means that we are observing a process in which $\Delta v = \pm 2$. This new selection rule can arise from two different causes: the mechanical anharmonicities of the harmonic oscillator or a non-linear term in the polarizability, i.e., $\frac{\partial^2 \alpha}{\partial Q^2} \neq 0$. In either case we have a sharp line corresponding to $\Delta v = \pm 2$ with the Raman displacement in general being a little less than twice the Raman displacements for the $\Delta v = \pm 1$ transition.

Raman scattering by phonons

The main difference between vibrational Raman effect in liquids and in solids is that in liquids we are scattering light by the changes of polarizability associated with a normal mode in the molecule. In solids, such as NaCl, if we make a pair $Na^+ - Cl^-$ oscillate, the sodium is tightly bound to all its Cl^- nearest neighbors so the change of position of this Na^+ ion will induce a change of position of all those Cl^- which, in turn, will induce changes in position of the next Na^+ ions, etc. In the crystal, the molecule NaCl loses its identity and when a certain NaCl "molecule" vibrates this vibration becomes a wave that goes through the crystal. This is a phonon, which is characterized by a certain phase velocity v, a frequency ω and a wave propagation vector k. The phonon is a normal mode of the crystal. In effect we can consider the Raman effect in liquids as one in which $|k| = 0$, i.e., there is no propagation.

So, while in liquids we have only to conserve energy in the Raman scattering process, in solids we have to conserve both energy and momentum or more properly, energy and wave vector k. Another difference which is very important between Raman spectroscopy of liquids and solids is that while the liquid molecules are randomly oriented in relation to the laboratory system of reference all the unit cells of a solid are oriented in the same manner. There is no difference, in the solids, between the laboratory and the crystal systems of reference. Imagine that we calculate from group theory that a primitive cell of a solid has

65

a vibration in which the only changing components in the polarizability tensor are xy and yx. We see immediately that to observe the Raman effect of that phonon, we should make the polarization of the incoming laser radiation parallel to the x axis and observe the light that is scattered with its polarization in the y direction or vice-versa, since only those two geometries would give non zero results for the equation

$$P = \alpha E \quad \text{i.e.} \quad P_x = \alpha_{xy} E_y \text{ and } P_y = \alpha_{yx} E_x \ . \tag{9}$$

So in very elegant ways, we can for a solid, determine all the Raman active phonons and to each we can associate a polarizability tensor and a definite symmetry. Figure 3 shows a typical example for MnF_2. MnF_2 belongs to the D_{4h} point group and group theory predicts four Raman active modes: $1A_{1g}$ (with α_{xx}, α_{yy}, $\alpha_{zz} \neq 0$), $1E_g (\alpha_{yz}$, α_{xz}, α_{zy}, $\alpha_{zx} \neq 0)$, $1 B_1 (\alpha_{xx}$, $\alpha_{yy} \neq 0)$ and one $B_{2g} (\alpha_{xy}$, $\alpha_{yx} \neq 0)$. Figure 3 shows the spectra observed for the different α_{ij} spectra, in complete agreement with the results of group theory [14].

The conservation of momentum plays a very important role in the understanding of the spectra of solids. First the frequency of a phonon varies with momentum throughout the Brillouin zone (the maximum $k = 2\pi/a$ where a is the crystal lattice constant). For example, Figure 4 shows how the frequency varies with momentum for two acoustical modes and two optical modes. Since we are using visible light, where $|k| = \nu/c \approx 10^5 cm^{-1}$ a 90° scattering will create or destroy phonons with $|k| \approx 10^5$ cm^{-1}; in the scale of Figure 4 this $|k|$ is very small so we often refer to the Raman effect measuring the $|k| \cong 0$ phonons. Let us see the influence of this conservation of momentum in another case. In ZnO there is a doubly degenerate phonon of symmetry E_1 which is both Raman and infrared active. When this phonon is infrared active in the x direction its polarizability tensor components xz and zx are different from zero and when the phonon is polarized in the y direction the yz and zy polarizability components are different from zero[15]. Imagine that we are looking at the xz polarizability component of this E_1 line in ZnO: if the light is incident in the z direction and the observation is made along the y direction we produce, by conservation of momentum, a phonon in the yz plane with x polarization (since we are measuring the xz component of the tensor). So this phonon has a propagation direction perpendicular to its polarization and is a transverse optical phonon (TO). If, still measuring the xz spectrum, the light is incident in the z direction and the observation is in the x direction we produce a phonon in the xz plane with x polarization. This observed phonon can be transverse (propagation perpendicular to polarization) or longitudinal (LO). Since the transverse and longitudinal have different frequencies we observe two lines, so in the z(xz)y spectrum we see only one line, the TO phonon, while in the z(xz)x spectrum we see two E_1 lines, the TO and LO (transverse optical and longitudinal optical) phonons. Figures 5A and 5B show the z(xz)x and the z(xz)y spectra of ZnO with the conservation of momentum diagrams which explains them[16].

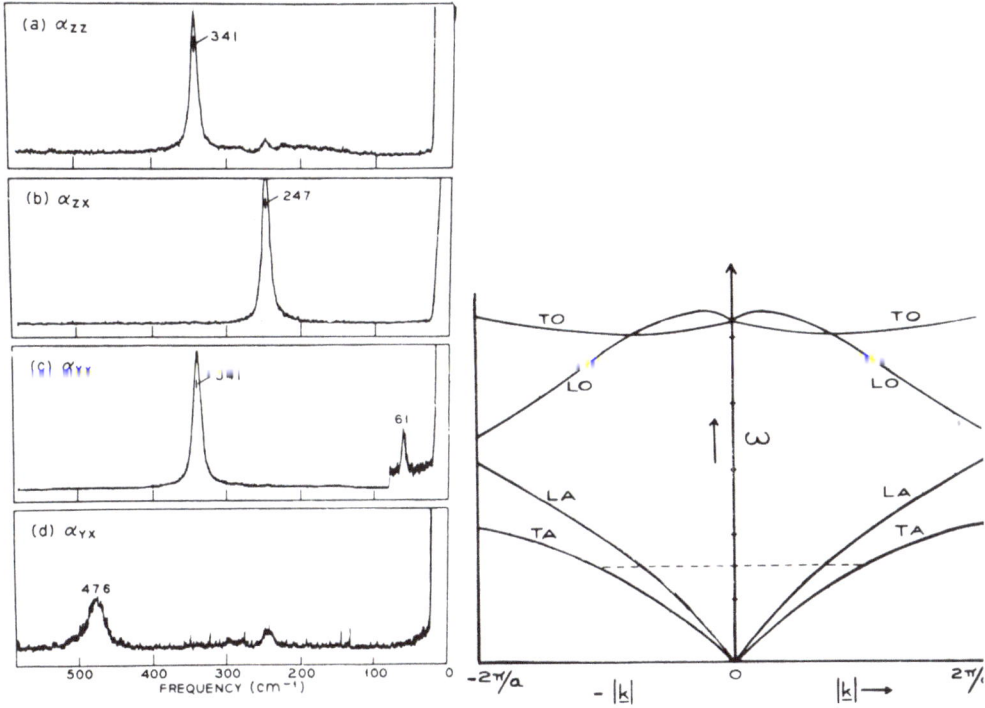

FIGURE 3. Raman scattering of MnF_2. In this crystal there is an A_{1g} vibration at 341 cm^{-1} with $\alpha_{xx,yy,zz} \neq 0$, a B_{1g} vibration, with $\alpha_{xx,-yy} \neq 0$ at 61 cm^{-1} a $B2g$ line with $\alpha_{xy,yx} \neq 0$ at 476 cm^{-1} and an Eg vibration at 247 cm^{-1} with the $\alpha_{xz,yz}$ 0 (From [14]).

FIGURE 4. Idealized dispersion curves of the acoustical and optical phonons of a crystal with two atoms per unit cell showing how the phonon frequencies change with wave vector

Raman scattering by multiple phonon processes

If we expand the polarizability tensor as done in Eq. (8) and keep higher order terms:

$$\alpha = \alpha_0 + \frac{\partial \alpha}{\partial q_i} q_i + \frac{\partial^2 \alpha}{\partial q \partial q_j} q_i q_j . \tag{10}$$

We see that the production of two or more phonons will create a modulation in the polarizability and will scatter in a Raman-like process. Similarly, if the force constants between atoms are not that of harmonic oscillators but have terms like ax^3, bx^4, etc., the selection rules are relaxed and $\triangle n = \pm 2$, ± 3 processes are now allowed.

Again, this scattering process has to conserve both energy and momentum; for a two phonon process:

$$\omega_L = \omega_R \pm \omega_{ph1} \pm \omega_{ph2}, \quad k_L = \bar{k}_R + \bar{k}_{ph1} + \bar{k}_{ph2} \cdot \tag{11}$$

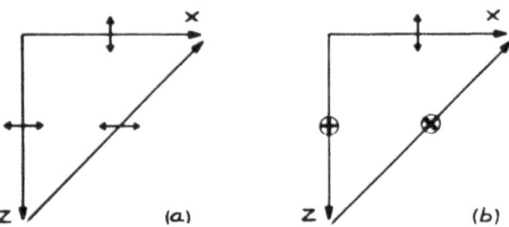

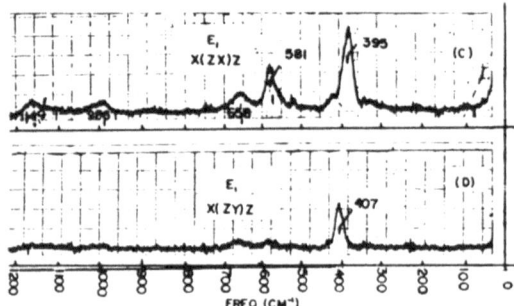

FIGURE 5. Experimental demonstration of the conservation of momentum in an optical phonon scattering in ZnO. Figure 5A is the conservation diagram to show that in 5C one obtains the scattering from LO and TO phonons. Figure 5B explains the data of Figure 5D where the scattering is obtained from the TO phonon only. (From [16]).

If, as in the case of liquids, only phonons with $k=0$ exist we have the production of an overtone Raman effect which as indicated before consists of sharp lines. In the case of two phonons scattering in solids, the two phonons can be produced throughout the whole Brillouin zone with the only conditions that energy and momentum are conserved in the scattering process and that the process has the correct polarizability tensor of the experiment. Since, as seen in Figure 4, the frequency of a phonon varies drastically with momentum, in general multiple phonon scattering is very broad in frequency. We can always imagine two successive Raman processes, i.e., one phonon is produced, the Raman light is then scattered again, and another phonon is produced which will produce a sharp two phonon or overtone line. In general, however, the two phonon processes are broad even though most of its intensity come from large $|k|$ phonons close to the edge of the Brillouin zone, because the density of states at the critical points is in general larger than that for phonons with small $|k|$.

Besides the broadness which sometimes can be misleading, there are two other ways of recognizing if a process is a multiple Raman scattering. First, group theory will tell you that the symmetry of the two phonon process is the product of the symmetries of the two phonons involved in the process. In general then, the symmetry of the two phonon process is more complex than that of the one phonon, and sometimes would appear not to obey the symmetries allowed for the specific point group under study. Since the product of two symmetries quite often has the most symmetric representation, A_1, in it, the second order Raman processes quite often appear together with the A_1 spectra. Another way to recognize a second order Raman process is its temperature dependence. The one phonon Stokes process decrease in density with temperature as $(n+1)$ while the two phonon process occurring at the same frequency will have a variation of intensity with temperature obeying a law like $(n+1)^2$ where $n = (\exp hW/KT -1)^{-1}$. The two phonon process then fades away quickly with the decrease of temperature.

The old concept that in the Raman processes the strong lines are one phonon lines while the two phonons are broad and weak, is very treacherous. We know of many instances like $BaTiO_3$, TiO_2, $KTaO_3$, etc., where some of the most prominent features of the Raman spectra are due to multiple phonon processes. Another fact to remember is that the $|k| \simeq 10^5$ acoustical phonon gives rise to the Brillouin spectrum with frequency of the order of $.1$ cm^{-1}, while the two phonon Raman scattering of the acoustical processes will extend to a few hundred of cm^{-1}, since most of the Raman effect is due to phonons with $|k|$ near the edge of the Brillouin zone.

Figure 6 shows part of the (xx) spectrum of TiO_2 where the sharp line at 143 cm^{-1} corresponds to a one phonon process of symmetry $B_1(\alpha xx, yy \neq 0)$ while the broad and strong band at $\sim$234 cm^{-1} corresponds to the two acoustical phonon Raman scattering [14].

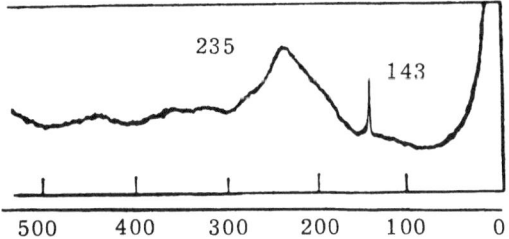

FIGURE 6. Part of the xx spectrum of TiO_2 showing that the two phonon process at $\sim$235 cm^{-1} can be stronger than the one phonon B1g scattering at 143 cm^{-1}. (From [14]).

In principle, the two phonon process should give us considerable information on the dispersion relation ($|k|$ vs ω) of phonons, critical points in the Brillouin zone where the population of those phonons is maximum, etc. However, due in part to the poor state of the experimental work on the second order Raman processes, we feel that most of the original promises have not been fulfilled.

Raman effect of F centers and impurities in crystals

Imagine that we have a crystal like NaCl in which the first order Raman effect is forbidden for reasons of symmetry (each Na and Cl ion sits at a center of the cubic lattice). If we remove a Cl atom and substitute it by a vacancy which traps an electron, we create an f-center. By creating a f-center we destroyed the translational symmetry of the crystal so that the Na^+ ions which are next-neighbors of the trapped electron are no longer at centers of cubes. First order Raman, i.e., one phonon processes, are now allowed around the f-center. The same happens, for instance, when we substitute a Cl by a Br or a Na by a K atom, for example.

The problem of observation of Raman effect of f-centers is that we cannot introduce too many of them and by just having a few of them the Raman effect is weak. In the case of NaCl, for instance, with 10^{17} f-centers/c.c. the first order Raman effect intensity of f-centers at 300°K is at best comparable to the weak two phonon spectrum of the crystal. To observe the scattering from those f-centers the temperature is lowered, to discriminate against the two phonon processes and the frequency of the laser excitation can be chosen to be close to the strong electronic absorption of the f-center so that the resonant denominator will increase the cross section of the Raman process [17]. An impurity Raman effect in alkalihalides, for instance, cannot make ready use of this resonant denominator because the visible absorption of the impure crystal will not change drastically. Much larger concentration of the impurity can be used without appreciably disturbing the crystal symmetry and the total cross section can be then made sufficiently large for observation [18].

An interesting characteristic of both the f-center and impurity Raman effect is that the Raman active centers, or impurities, have lost the translational symmetry which is characteristic of a solid, so that conservation of momentum loses its meaning and the $|k|$ of the phonon can, in essence, have any value within the Brillouin zone. This means that, like in the two phonon processes, the scattering will include phonons with all $|k|$'s and the spectrum will be broad reflecting the density of states functions for all the allowed phonons, instead of the normally sharp one phonon processes obtained for a solid where one is sampling only the $|k| \simeq 0$ phonons.

Raman effect of polaritons

Electromagnetic radiation passing through a crystal is characterized by a frequency ω, a velocity (c/n) and a wave vector k. For low frequency light, the dispersion relation (ω vs. k) is a straight line passing through the origin. Imagine that we plot in the same graph the dispersion relation (ω vs. $|k|$) of light and of an infrared active optical phonon with its LO and TO components; the k interval we are interested in is so small that we may consider that the phonon frequency itself is constant independent of $|k|$ As seen from the dashed lines of Figure 7 the two dispersion relations will cross, meaning that at the crossing the phonon and the electromagnetic radiation will have the same frequency and wave vector. If this phonon is infrared active there will be an interaction of the electromagnetic radiation and mechanical vibration and the excitation, around the interaction region, will be partially phonon and partially light. This mixed excitation in the interaction region is the

the polariton and its dispersion relation is also shown in Figure 7 as the full lines. As it can be seen from the figure we have two branches of the polariton: the upper or quasi-photon branch ω^+, which in the limit $|k| \to 0$ tends to the frequency of the LO mode and which has escaped observation up to date and the lower or quasi-phonon $\omega-$ which has been observed for GaP [19], ZnO [20] and quartz [21]. From Figure 7 we see that the polariton exists only for very small values of $|k|$ so to scatter from it we have to observe the Raman effect, effectively in the forward region. Since this scattering process conserves momentum, in the forward direction we will produce its minimum $|k|$ and by observing the scattering let us say at 1°, 2°, 3°, etc., from the forward region we observe polariton of higher and higher $|k|$ until the excitation becomes pure phonon for angles of the

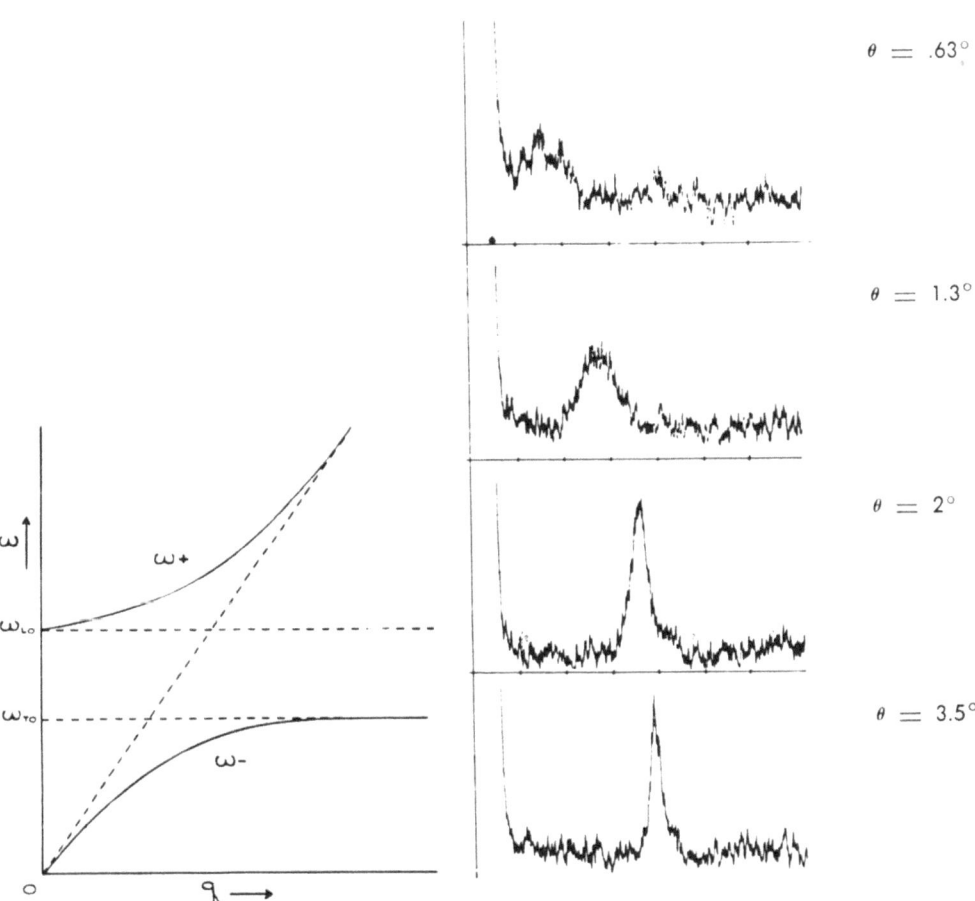

FIGURE 7. Idealized dispersion curves of the coupled phonon — E. M. wave system (polariton) for the case of one infrared and Raman active phonon interacting with long wavelength light. The dashed lines are the dispersion of the uncoupled phonon and light waves. The solid curves are the dispersion of the coupled excitations.

FIGURE 8. Experimental observation of the scattering of polaritons. One can see easily that the frequency shift changes with the angle of observation from the forward direction. At 3.4° from the forward direction the polariton is almost completely the TO phonon whose frequency is 407 cm^{-1}. (From [20]).

71

order of 10°. Figure 8 shows the Raman effect of the polariton in ZnO [20], in which observation of the dispersion relation of quasi-phonon polariton from 160 to 407 cm^{-1} is shown.

Raman effect of spin waves or magnons

Spin waves are excitations characterized by dispersion relations (ω vs. $|k|$) very much like those of phonons. They occur in magnetic materials which have atoms with the non-zero spins oriented in an ordered manner. A ferromagnet (such as iron, nickel, etc.) has all the atoms with spins which are parallel and with the same orientation giving rise to very high degree of magnetization. An antiferromagnet (MnF_2, FeF_2) has one of the magnetic ions pointing the spin in a definite direction while the next ion points the spin exactly in the opposite direction, etc.; the total magnetization of an antiferromagnet is zero but still all the spins are oriented. In a ferrimagnet the spins of next neighbors are also antiparallel like in an antiferromagnet but they are of different magnitude so there will be no cancellation and a magnetization remains, see Figure 9. Exactly as in the case of phonons, if we now disturb the orientation of one spin, since they are all coupled, this misorientation will be felt by the next neighbors, etc., creating again a wave which will go through all the spins. This process, or this wave, is called a spin wave or a magnon.

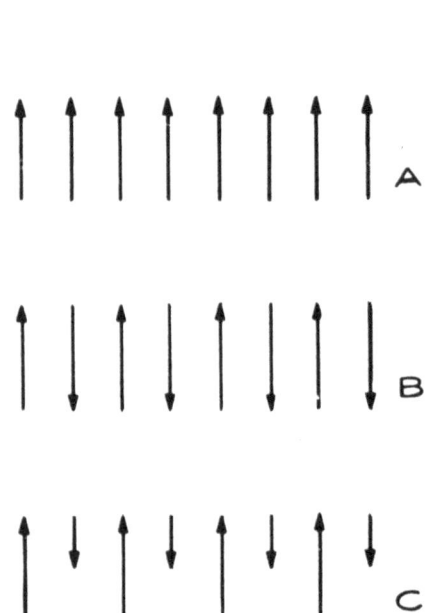

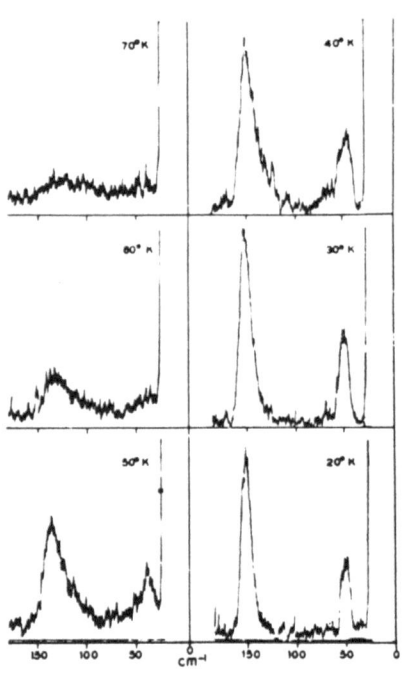

FIGURE 9. Graphical representation of the spin alignment of (a) ferromagnetic materials, (b) antiferromagnets and (c) of ferrimagnets.

FIGURE 10. The scattering from the one and two magnon states of FeF_2. FeF_2 becomes antiferromagnetic at ~70°K so the spins become oriented at that temperature and spin waves can propagate and scatter light. (From [22]).

72

If we heat a magnetic material we can break up the spin ordering and destroy its magnetic properties. When cooling the sample, the temperature in which magnetic ordering takes place is called the Neel temperature. In Figure 3 we see the room temperature spectrum of MnF_2 without any trace of spin waves scattering; by cooling the crystal below its Neel temperature ($\sim 70°K$) Raman scattering by spin waves show up [22]. Figure 10 shows the dramatic appearance of the one and two magnon processes in FeF_2 [22] as the temperature is lowered.

As in the case of phonons, the dispersion relations of magnons are not flat so the frequency of the broad two magnon scattering does not have to occur at twice the frequency of the one magnon process since the two magnon process consists of scattering by a pair of magnons throughout the Brillouin zone. The frequency distribution of the two magnon processes, as in the case of phonons, reflect the dispersion relation of the magnons in question and their density of states for the different points in the Brillouin zone [23].

Electronic Raman effect

Electronic Raman effect is the light scattering by a material in which conservation of energy is furnished by the quantum jump of an electron from one electronic state to another. This effect is an old one, as far as Raman effects go, since it was discussed by Placzek in 1934, who also discussed the experiments done up to that time [7]. More recently the electronic Raman effect in solids was observed with mercury excitation in which the electronic transitions were between the Stark split levels of a rare earth ion in a crystal field [24] and even more recently electronic Raman transitions were observed between the ground and impurity level, in semi-conductors [25].

It is interesting to discuss what new information one might obtain from the Raman spectrum of rare earth Stark split levels. Let us imagine that we dope a crystal of $LaCl_3$ with Pr^{+3}. The crystal field of $LaCl_3$ will split the degeneracy of all the degenerate Pr^{+3} levels. The ground state of $Pr^{+3}(^3H_4)$ for instance will be split into five levels each one of which can be characterized by a wave function ψ, an energy E, a quantum number μ and a symmetry S in relation to all the operations of the crystal. So the electronic Raman effect of the Stark split levels of the $^3H_4$ state of Pr^{+3} will consist of four lines which will be characterized by Raman displacements ΔE and polarizability tensors for each of the electronic transitions. These electronic Raman effect tensors, in general, will be more complicated than those obtained for the phonons in the same crystal but at the same time they will contain more information. From the form of the tensors for the different transitions we should be able to obtain many correlations such as energy levels with their quantum numbers and wave functions.

One might question the usefulness of the electronic Raman effect since most of the information can also be obtained from absorption and fluorescence spectroscopy. The answer is the same as can be given for phonons: in a dipole absorption or fluorescence we measure the three possible components of the dipole vector, in Raman we measure the nine components of a tensor and inherently we should get much more information from the tensor.

Plasmon scattering

Imagine that we have N electrons inside a cube of volume A^3. There will be coulombic repulsions between the electrons and they will tend to organize themselves around configurations of minimum free energy. If this box of electrons is cooled to 0° K the electrons would organize themselves like atoms in a crystal. If now in this configuration one moves one electron around its equilibrium position, this disturbance will propagate itself through the medium as a wave which we call a plasmon wave.

The plasmon, like the other excitations studied, can be characterized by its frequency ω_p and by its momentum k which are given by:

$$\omega_p = \left(\frac{2\pi\rho\ell^2}{Em*}\right)^{\frac{1}{2}} \quad \text{and} \quad |k| = \left(\frac{4\pi\rho\ell^2}{KT}\right)^{\frac{1}{2}} , \qquad (12)$$

where ε is again the dielectric constant, ρ is the plasma density and m* is the effective mass of the electrically charged particle. For a gas and a metal $\varepsilon \sim 1$ and for a semiconductor $\varepsilon \sim 10$ so for metals $\omega \sim 10^6 cm^{-1}$, for a gas plasma $\omega \sim 10^1$-$10^4 cm^{-1}$ and for a semiconductor plasma $\omega \sim 10^2 - 10^7 cm^{-1}$.

Let us scatter a laser beam from the plasma. We are going to transfer momentum in the scattering process, from the light wave to the plasma: if the scattering experiment is done for instance at 90° the momentum transferred is $\sim 10^5 cm^{-1}$ and if the scattering is observed in the forward direction the transferred momentum $\underline{k}$ goes to zero. The important thing in the scattering process is that the momentum transferred has to be transferred to the plasmon and we have to have allowed plasmons with the needed $|k|$. If the $|k|$ to be transferred in a scattering experiment is larger than the allowed $|k|$ for the plasmon, we have no scattering from those plasma waves.

We can imagine qualitatively that in a scattering process the $|k|$ transfer measures the coarseness of a system. Large $|k|$ transferred means large frequencies or small wave length intervals probed. If the collective phenomena like a gas plasma involves particles far away from each other its frequency and momentum (wave-vector) are small, so in order to see it we have to probe it with a small momentum transfer.

If the momentum transfer is smaller than the maximum allowed momentum of the plasmon in a certain system we can see the collective or plasma excitations, but if the momentum transfer is large, we are probing smaller volumes or in a plasma we are probing the velocity distribution, for instance, of the individual scattering charges. This is beautifully illustrated in Figures 11a and 11b [26] which shows the near forward and 90° scattering of ruby laser light by a flash produced H_2 plasma.

In the case of Figure 11a we see the Rayleigh and plasma shifted frequencies. This scattering is essentially similar to the Brillouin scattering where the Rayleigh and plasma frequencies can be understood as propagating (pressure) and nonpropagating (entropy) fluctuations in the density of charges; the plasma shift observed is a function of $|k|$ in similar way as a Brillouin experiment. In Figure 11b we obtain the scattering regime for high $|k|$ where in essence we see the Doppler broadened scattering by the individual particles, in this case molecular or atomic hydrogen ions surrounded by an electron cloud.

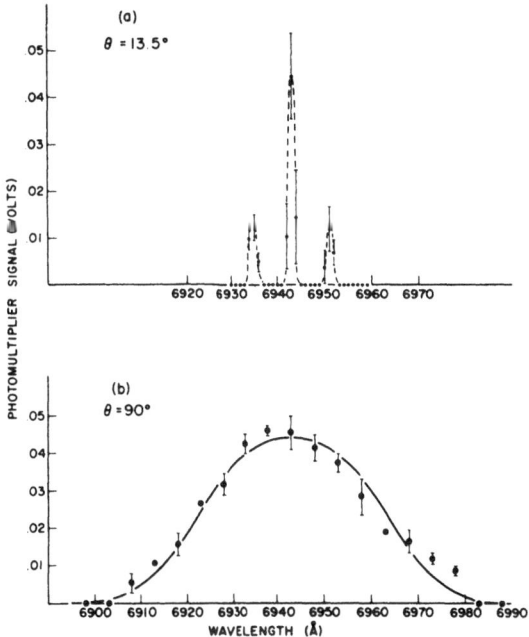

FIGURE 11. Light scattering from a gaseous hydrogen plasma at the ruby laser frequency. On 11(a) the scattering was observed at small angles and the plasma as well as the Rayleigh scattering are observed. At large angle, 90°, and large $|k|$ transfer, we see the Doppler broadened scattering of the elementary scattering particles and not the collective excitations. (From [26]).

We have treated above only "free" plasmas, not coupled to the medium through which they flow. Let us now consider this plasma in a solid, let us say a semiconductor. We still have the same "Brillouin-like" spectrum discussed above with the electronic mass m substituted by an effective mass m*, but also there will be a coupling of the lattice with the oscillating plasma electric field. An external electric field couples to the lattice and is in effect responsible for the LO-TO splitting of optical vibrations, and this coupling is felt also in the dielectric constant of the material. While in a plasma-free medium we can write that the dielectric constant ε at the low frequencies is

$$ E(\omega) = E_\infty + \sum_i \frac{\omega_i^2 S_i}{(\omega_i^2 - \omega^2)} , $$

where ε_∞ is the dielectric constant far away from the optical phonon frequencies, ω_i is the frequency of the infrared active optical phonons (transverse, or T O modes) and S_i is the intensity of each i mode. If we now add free carriers with a plasma frequency ω_p given by (12) the dielectric constant expression has to be modified to

$$E\ (\omega) = E_\infty + \sum_i \frac{\omega_i^2\ S_i}{(\omega_i^2 - \omega^2)} - \frac{\omega_i^2\ E_\infty}{2} \tag{14}$$

We know that the infinities of the dielectric constant give us the frequencies of the T O , or transverse optical modes and from expression (14) we see that the plasma does not change these T O frequencies, so we can say the plasma does not couple to the T O modes. The zeros of the dielectric constant are the L O (longitudinal) mode frequencies and those are definitely modified by the plasma. As a matter of fact we have now for each T O mode two coupled L O − plasmon modes with their frequencies, line widths, etc., completely dependent on the plasma frequencies.

This coupling plasmon - longitudinal mode - lattice we can classify as a generalized Raman effect. Figure 12 shows the Raman effect of GaAs [27] with different carrier concentration in which the T O mode appearance is hardly affected while the LO-plasma modes are very dependent on carrier concentration.

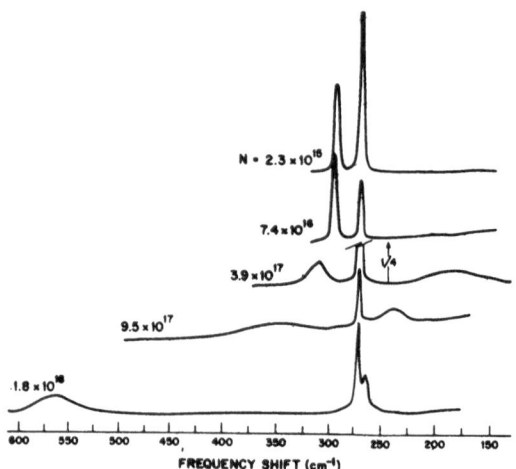

FIGURE 12. Scattering from coupled LO-plasma modes in GaAs, from Ref. [27]. On top for a low carrier concentration we see the LO and TO modes of GaAs. As the concentration of carriers N is increased there is a coupling of the LO and plasma frequencies shown very clearly.

Raman scattering of Landau levels

If we subject a free electron to a magnetic field it will describe a circular trajectory with the radius and frequency determined by the magnetic field, electronic charge, mass of the particle and the dielectric constant of the medium. Those "cyclotron orbits" inside a crystal are quantized and the quantized levels are called Landau Levels. If we choose a crystal in which the effective electronic mass is small, for example in GaAs the effective electron mass is ~.07 and in InSb it is ~.01 of the rest electron mass, we can make a large Landau splitting for relatively small magnetic fields.

The Landau levels of an electron in a magnetic field are equally spaced in energy, reflecting the fact that one can consider them as levels of an harmonic oscillator in which the fundamental frequency is equal to the cyclotron frequency ω_c. If we carry the harmonic oscillator approximation further we should calculate the Raman selection rules and cross section for a harmonic oscillator and the selection rules would be $\Delta n = 0, \pm 2$ [28]. If we include anharmonic terms $\Delta n = \pm 1$ would be allowed but should naturally be weaker than the $\Delta n = \pm 2$ transitions. Recent experiments [29] in Raman scattering by Landau levels in InSb were performed and in first order we can say that theoretical predictions of the properties of the scattering were observed. The $\Delta n = \pm 1$ transitions, however, were allowed in the same order as those with $\Delta n = \pm 2$ so the harmonic oscillator approximation for the Landau levels has to be modified.

Electric field induced Raman effect

If we apply an electric field to a collection of coupled harmonic oscillators we can induce a dipole proportional to the polarizability, and vibrations which were only Raman active without the field become infrared active. This is the well known effect of Stark induced infrared absorption. One can also show that the electric field will shift all the levels of the harmonic oscillator by the same amount, so the transition frequency does not change nor do the Raman effect $\Delta n = \pm 1$ selection rules for the oscillator change. The electric field may however change the symmetry of the unit cell in such a way that, for instance, it may pull the center atom of a cubic crystal of symmetry O_h off the center position making the symmetry of the cell a tetragonal C_{4v}, for example.

This small displacement of an atom in the unit cell may have drastic effects. Imagine that for reasons of symmetry a crystal has no first order Raman effect: the electric field by so slightly changing the symmetry makes the first order Raman effect allowed without affecting, in first order, the vibrational frequencies of the field free crystal.

If the change of symmetry is small the new one phonon Raman effect at first will be as strong as the two phonon process until hopefully it will become so strong that it will dominate the spectrum. If one has the case in which the new, induced, one phonon process is only as strong as the no field allowed, two phonon process, there is a clever way of separating the two [30]. The laser is continuously shining upon the crystal while the electric field is a square wave of frequency ω, let us say 100 cycles/sec. The detection of the Raman effect is made in a synchronous amplifier set at the electric field frequency ω or at 2ω; the electronics are then measuring the difference between the field-on and field-off spectra and while the two phonon spectra cancels out, the newly allowed field induced one phonon process stands out.

77

This technique might allow a new look at the forbidden modes and at a host of new crystals while per se the effect will give information on the effect of electric fields on vibrational mode frequencies, etc.

CONCLUSION

We have tried to give a bird's eye view of what is going on today in the field of light scattering with laser sources. So much is going on in fact that all of those working in the area wait eagerly for each number of the Physics Review Letters to see what new effects are described or which unexplained details of the older ones have been clarified. It is an age of excitement in the light scattering field, brought about by the availability of the laser sources.

REFERENCES

1. LORD RAYLEIGH, Phil. Mag., **47**, 375 (1899), Coll. papers IV, p. 397
2. L. BRILLOUIN, Ann. Phys., **17**, 88 (1922)
3. A. SMEKAL, Naturwiss., **11**, 873 (1923)
4. C. V. RAMAN, Indian J. Phys., **2**, 387 (1928); C. V. RAMAN and K. S. KRISHNAN, Nature, **121**, 501 (1928)
5. G. LANDSBERG and L. MANDELSTAM, Naturwiss., **16**, 57 (1928)
6. E. GROSS, Nature, **126**, 201, 400, 603 (1930)
7. G. PLACZEK, Marx's Handbuck der Radiologie VI, **2**, 209 (1934)
8. In particular the Indian work published in the Ind. Acad. Sc. by R. S. Krishnan and others and in this country by D. H. Rank et al., and more recently with laser sources by R. Y. Chiao and B. Stoicheff, J. opt. Soc. Am., 54, 1286 (1964)
9. G. BENEDEK, J. B. LASTOVKA and K. FRITCH, J. opt. Soc. Am., **54**, 1284 (1964); G. MINICHINO, R. O'BRIEN, G. J. ROSASEO and A. WEBER, Bull. Am. Phys. Soc., **12**, 1132 (1967). Both groups used laser sources.
10. J. B. LASTOVKA and G. B. BENEDEK, Proc. Phys. Quantum Electr. Conf., Puerto Rico, p. 231, (1966), M. G. COHEN and E. I. GORDON, Bell Syst. Tech. J., **44**, 693 (1965) and H. Z. CUMMINS, N. KNABLE and Y. YEH, Phys. Rev. Letters, **12**, 150 (1964)
11. A. WEBER, S. P. S. PORTO, L. E. CHEESMAN and J. J. BARRETT, J. opt. Soc. Am., **57**, 19 (1967)
12. P. DEBYE, Polar Molecules, Ch. V, Dover Publ., N. Y. (1929)
13. S. L. SHAPIRO and H. P. BROIDA, Phys. Rev., **154**, 129 (1967)
14. S. P. S. PORTO, P. A. FLEURY and T. C. DAMEN, Phys. Rev., **154**, 522 (1967)
15. R. LOUDON, Adv. Phys., **13**, 423 (1964)
16. T. C. DAMEN, S. P. S. PORTO and B. TELL, Phys. Rev., **142**, 570 (1966)
17. J. M. WORLOCK and S. P. S. PORTO, Phys. Rev. Letters, **15**, 697 (1965)
18. J. HURRELL, T. C. DAMEN, S. P. S. PORTO and S. MASCARENHAS, Phys. Letters, **26A**, 194 (1968)
19. C. H. HENRY and J. J. HOPFIELD, Phys. Rev. Letters, **15**, 964 (1965)
20. S. P. S. PORTO, B. TELL and T. C. DAMEN, Phys. Rev. Letters, **16**, 450 (1966)
21. J. P. SCOTT, L. E. CHEESMAN and S. P. S. PORTO, Phys. Rev., **162**, 834 (1967)
22. P. A. FLEURY, S. P. S. PORTO, L. E. CHEESMAN and H. J. GUGGENHEIN, Phys. Rev. Letters, **17**, 84 (1966)
23. P. A. FLEURY, S. P. S. PORTO and R. LOUDON, Phys. Rev. Letters, **18**, 658 (1967)
24. J. T. HOUGEN and S. SINGH, Phys. Rev. Letters, **10**, 406 (1963)
25. C. H. HENRY, J. J. HOPFIELD and L. C. LUTHER, Phys. Rev. Letters, **17**, 1178 (1966)
26. S. RAMSDEN and W. DAVIES, Phys. Rev. Letters, **16**, 303 (1966)
27. A. MOORADIAN and G. B. WRIGHT, Phys. Rev. Letters, **16**, 999 (1966)
28. P. A. WOLFF, Phys. Rev. Letters, **16**, 225 (1966)
29. R. E. SLUSCHER, C. K. N. PATEL and P. A. FLEURY, Phys. Rev. Letters, **18**, 77 (1967)
30. J. M. WORLOCK and P. A. FLEURY, Phys. Rev. Letters.

4 / Vision: Human and Electronic

A. ROSE

I. Preface

The series of lectures on the process of vision in both human and electronic systems was based predominantly on a number of publications in scattered parts of the literature. Several of these papers are reproduced here and serve, at least, the convenience of juxtaposition.

A scanning of the titles of the lectures may suggest a disjointed collection of topics. All of this is true to the extent that the topics are not likely to be found in any one journal or text or conference. The common thread, however, can be readily identified. A perfect system for vision is one that, in effect, counts or senses each incident photon. In this way all of the information in the incoming stream of photons is reproduced. The information in the incoming stream itself is limited both by the finite number of photons and by the fact that they are randomly spaced in time. The latter gives rise to a fluctuation or noisy background against which the picture is viewed. This source of noise is fundamental and, in the perfect system, should be the dominant source. It is relatively easy to design near perfect systems for vision at high light intensity where the level of the fundamental photon noise is correspondingly high and can readily dominate local sources of noise in the receiving system. The major problems arise as one attempts to see at extremely low levels of illumination. In the limit, the ability to detect individual photons as individual events is a necessary condition for a perfect system for vision. The series of lectures is concerned with how well the human system succeeds as a photon counter and with how well various electronic devices have been or can be designed to be photon counters.

The first paper on "Quantum Effects in Human Vision" outlines quantitatively how much information is contained in a given number of photons and compares this to the information actually transmitted by the human visual process. The conclusion is not only that the human eye is a remarkably sensitive device, approximating 10% of a perfect device, but, even more to the point, that the high performance is maintained over a range of about 100 million fold in light intensity.

The second paper on "Television Camera Tubes and the Process of Vision" compares the performance of a number of electronic pick up devices with that of the human eye and also with that of a perfect device. While various camera tubes rival the human eye at intermediate light levels and at high lights, no one device at the time the paper was written covered the full range of light levels covered by the eye. Recently, various forms of image intensifier arrangements have been combined with image orthicons to extend the range of electronic vision to light levels even below the absolute threshold of the eye. Descriptions of the more recent tubes are to be found in the review paper by P. K. Weimer cited in the text.

The television camera tubes developed up to the present have been vacuum tube devices in which the target is scanned by an electron beam. Recently, self scanned arrays of solid state photoelements have been proposed and demonstrated. These have the virtue of compactness and the possibility of high sensitivity. The latter half of the lecture series is concerned with the physics and electronics which bear on the question of whether a solid state photoelement can detect individual photons. The papers on the Gain-Bandwidth Products for Photoconductors and Solid State Triodes* lay the groundwork for the calculation of the conditions to be satisfied if a photoconductor is to count individual photons. The groundwork is largely an analysis of space-charge-limited currents in solids. The major conclusion here is that a photon counter is possible provided the thermal noise in the photoconductor is not amplified by the applied fields required to achieve a photoconductive gain of 10^3 to 10^4.

Another formal method of achieving a photon counter in a solid state diode is by way of a solid state photomultiplier. The original photo-excited electron generates a thousand or more descendants by impact ionization. High electric fields in the neighborhood of dielectric breakdown are generally needed in order to heat up the electrons to the point of impact ionization. The subject of hot electrons is intimately related to the more general subject of electron-phonon interactions and to the range of validity of ohm's law. Treatments of electron-phonon interactions have been almost exclusively confined to the formalism of perturbation theory and are not designed for the practicing engineer. The last paper on "Hot Electrons and Ohm's Law" outlines an intermediate approach to electron-phonon interactions based mainly on the classical physics of polarization and carried out in real. as opposed to Fourier, space.

---

* Although presented as part of the lecture series, this subject was omitted from the text by an oversight on the part of the author. A version of it may be found in R C A R e v., 120, 157 (1959)

II. Quantum Effects in Human Vision

I. INTRODUCTION

The familiar story of forsaking one's gem-studded backyard to look for diamonds abroad takes on a particularly ironic twist in the case of the quantum theory of radiation. Planck made a magnificent excursion into radiation theory to establish the fact that radiation is not a smooth flowing continuum of energy but rather a finely divided stream of separable and countable bits of energy called quanta or photons. To the layman, Planck's demonstration was far from simple. It was a highly technical argument showing that only by the assumption of such discrete bits of energy could one account for the experimentally known spectrum of radiation from an incandescent body. Later, more direct and simple evidence for photons was supplied by the experiments on photoemission. In these experiments, the energies of electrons ejected from a solid by light were easily traced to the discrete energies of the photons that ejected them.

It is ironic that this same granular nature of light was all the while imposing its inescapable limitations on what the early scientists could literally see in their dimly lit libraries and laboratories. In fact, if the present writer is not deceived by his own observations, the early workers had only to glance up at some shadowed wall to see directly the granularity of the radiation they were intently investigating. This is not to suggest that acceptance of the quantum theory of radiation would have been hastened by pointing out its visual effects. Agreement comes more slowly in the field of vision than in the purely physical sciences.

Perhaps it is even more surprising, now that the quantum nature of radiation had been established for some 50 years, that it had not been used immediately to outline the absolute limits of visual performance. Only in the last few years has this limitation been seriously recognized, mainly by a small group of research workers whose business is to supplement and extend the range of human vision by electronic devices. The several attempts to see down to the fluctuation limits imposed by X-ray photons both in medical and industrial radiography are the most familiar current examples [4, 5, 14, 17-19, 28]. The somewhat longer standing problem of developing a television camera tube to match the human eye is another example [16, 24, 26, 31].

The problem of the quantum (or photon) limitations to vision can be quite simply stated. At any given light intensity, one counts up the total number of photons entering the eye during its storage time. From the total number of photons, one computes or constructs the best picture permitted by the random properties of the photons (Plate I). One then compares this picture with what one actually sees. If the two are the same, the conclusion must be that visual information is limited by the

Reprinted from: A d v. b i o l. m e d. P h y s., 5, 211—242 (1957). Academic Press Inc., N. Y.

finite number of photons entering the eye. This conclusion is absolute and is independent of one's choice of physiological or chemical mechanisms to be associated with the visual process.

If, on the other hand, this comparison shows that we actually see less than what is theoretically permitted, two conclusions are possible. The simplest would be to say that only a fraction of the stream of photons incident on the eye eventually contributes to forming our visual image. For example, it is known that half the light is reflected or absorbed in the inert lens and optical fluids before it reaches the retina. It is also reported that the sensitive retinal elements themselves, the rods and cones, absorb only a fraction of the light incident on them. This conclusion would still refer our visual information back to the finite stream of photons but would make use of only a fraction of that stream.

The second broad conclusion theoretically possible is that all of the light goes to forming a visual image but that there are local limitations imposed on what we can see. One kind of limitation could be of the threshold type in which a minimum number of photons must be accumulated before a sensation is triggered off. Much of our visual information would then be abortive. Another kind of limitation could come from what might generally be called "system noise." This is a familiar term in electrical communications used to describe the level of local disturbances in a receiver. An incoming signal must exceed this level before it can be detected. A more homely parallel is that the buzzing or ringing sounds one sometimes experiences in his ears obviously interfere with what he can hear. Possible evidence for local disturbances or "noise" in the visual system will be discussed later.

Whatever the local limitation postulated, one must take care that it does not deteriorate the theoretical picture too much. For example, if we actually see half of what is theoretically possible, the local limiting mechanism should not cause more than this factor of two deterioration. As obvious as this caution may appear, it is not trivial. A long standing model used to explain dark adaptation and identified by the phrase "bleaching of the visual purple," has maintained an apparently healthy existence in spite of the fact that its consequences are in patent violation of the limits set by the photon nature of light. In this sense, a knowledge of the absolute limits set by the photon nature of light can help to sift out various proposed visual mechanisms.

An actual comparison between what one sees and what one should theoretically see if he made use of every incident photon shows that our visual performance parallels closely the theoretical limits over a range of light intensity of 10^7. Even though our actual performance is a power of ten lower than the theoretical limit, the close approach to theory over this large range of light intensities strongly suggests that the finiteness of the incident stream of photons rather than some local mechanism in the eye controls our visual performance.

II. QUANTUM LIMITATIONS TO VISION

1. Limitations Imposed by Finite Number of Photons

Suppose we wish to construct a picture making use of small white dots all having the same size. Pictures formed in this way are common in display signs and in certain paintings executed by stippling. The minimum number of white dots required is evidently given by the number of picture elements to be resolved. Such a picture would, however, lack half tones, at least in areas of elemental size. If we wish to make a good picture, we might ask that each picture element be capable of showing a brightness difference from its neighboring elements of 1%. Now if we are in complete control of how we dispose the white dots, we would need only 100 dots per picture element. In this way, neighboring elements having 99 and 100 dots respectively would display the required 1% difference in brightness. The total number of dots required is then 100 times the number of picture elements. We ask now whether this requirement entails any serious limitation on the performance of the eye.

The final limit of resolution of the eye is set mechanically by the size of its receptor elements (the rods and cones). One of these elements is little more than a micron in diameter and subtends an angle of about one minute of arc at the eye lens. Let us consider a picture element containing a number of foveal cones and subtending an angle of two minutes at the eye lens. We can project this small angle out to the scene we are viewing and count how many photons enter the eye from the intercepted scene element. At a scene brightness of 10 ft.-L (foot-lamberts), a representative value for indoor lighting, the number of photons aimed at the picture element per second is 10^6. Since the storage or exposure time of the eye is 0.2 sec, the number of photons per picture time is 2×10^5. This number is large compared with the 100 photons required to reproduce a brightness difference of 1%. Accordingly, if we were free to dispose the incoming photons in a regular array, the number of photons available would be far in excess of that needed for an excellent picture. This would be true even for illuminations well below that of moonlight. In reality, however, we are not free to dispose of the photons as we wish. We must accept them as they come, and they come at randomly spaced intervals. This randomness makes a heavy demand on the number of photons needed to reproduce small brightness differences.

2. Limitations Imposed by the Random Character of Photons

It is well known and well established that an average number of random events has associated with it fluctuations from the average whose root-mean-square (rms) magnitude is equal to the square root of the average number. This relation applies equally to photons falling on the eye, electrons emitted from a hot body, rain drops falling on a pavement (if the rain drops are independent of each other), and to the random walks common in diffusion problems. The meaning of the relation becomes clear from a representative test. Consider a square foot of surface to be uniformly illuminated. The word uniformly means that the average number of photons falling on each square inch is the same over all the surface. Thus if we count the number of photons falling on each of 10 neighboring square inches during a time of one second, these numbers will not be the same. We can take an average of the 10 numbers and compare it with a similar average taken elsewhere on the surface. These two averages will be closely the same. The fractional discrepancy will be reduced if we make the count for a minute instead of a second and, in fact, will approach zero as the averaging time increases indefinitely.

Once having defined the average number of photons falling on each square inch per second, we can return to the set of 10 numbers giving the actual numbers of photons falling on each of the square inches in a time of one second. If we take the differences between each of these 10 numbers and the average, square each difference, average the squared differences, and take the square root of the average, this number will be the root-mean-square (rms) fluctuation. It will also be close to the square root of the average number of photons. If the operation were repeated many times and an average of these rms values taken, the rms value so obtained would approach even closer to the square root of the average number of photons.

The well known steps leading up to the rms fluctuation have been deliberately repeated here to emphasize that the concept is a mathematical measure of the fluctuations. It is a highly convenient measure. For example, in circuit theory the rms fluctuations in a steady current can be treated immediately as a current from which the fluctuation (noise) power can be computed. On the other hand (and this is the kernel of the present argument), the rms fluctuation does not immediately inform one how large a signal is needed to be distinguishable from random fluctuations. The signal is the difference in average brightness or number of photons between two neighboring areas. That difference or signal, as will be discussed shortly, is found from experiment to be detectable when it is about five times the rms fluctuation.

All of the necessary concepts are now at hand for recomputing how many photons per picture element are required to see a brightness difference of 1% between neighboring elements. In the absence of random fluctuations, it was immediately clear that only 100 photons were required. In the presence of random fluctuations, the number will be considerably larger.

What we ask is that the difference quantitatively in average numbers of photons between neighboring elements be 1% of the average number* and that the difference be five times the rms fluctuation associated with the average number. Let the average numbers be N_1 and N_2. Then, approximately, the two conditions are expressed by:

$$N_1 - N_2 = 10^{-2}N_1 = 5N_1^{1/2} \tag{1}$$

---

* Strictly one should take some kind of mean between N_1 and N_2 for the average number. It is not important here since N_1 and N_2 are nearly the same.

From Eq. (1) we get:

$$N_1 = 250,000 \text{ photons per picture element}$$

This is in striking contrast to the 100 photons previously computed in the absence of fluctuations and is close to the 2×10^5 photons available from a 10 ft.-L scene. In brief, the number of photons available to the eye at the comparatively high brightness of 10 ft.-L is just sufficient to produce an excellent picture. At lower scene brightnesses down to absolute threshold in the neighborhood of 10^{-6} ft.-L, the fluctuation-limited picture quality must steadily decrease.

The comparison between the actually perceived picture quality and the theoretical fluctuation-limited quality will be carried out in a later section. At this point it is useful to translate Eq. (1) into readily observable quantities.

3. Fundamental Relation: $BC^2 a^2 = Constant$

Eq. (1) of the previous section can be rewritten in terms of the number of photons (n) striking the retina per unit area and the element of area (s^2) under consideration:

$$(n_1 - n_2)s^2 = 5(n_1 s^2)^{1/2} \tag{2}$$

If Eq. (2) is squared and all of the variables brought to the left side, it may be written as

$$n_1[(n_1 - n_2)/n_1]^2 s^2 = \text{constant} \tag{3}$$

In this form we can readily replace the three factors with quantities proportional to them: n_1 with the average scene brightness B; $(n_1 - n_2)/n_1$ with the contrast C between neighboring elements; and s with the angular size a subtended by the test element under consideration. Eq. (3) then becomes the simple fundamental relation:

$$BC^2 a^2 = \text{constant} \tag{4}$$

The constant includes the storage time t of the eye, the threshold signal-to-noise ratio k, the diameter D of the pupil of the lens, a factor θ giving the fraction of photons incident on the eye that are eventually used in forming the picture and a factor depending on the units of B, C, and a. For B in foot-lamberts, C in percent contrast, and a in minutes of arc the constant is [23]:

$$\text{constant} = 5(k^2/D^2 t\theta) \times 10^{-3} \tag{5}$$

where D is measured in inches, t in seconds, and k has the value 5. In the example discussed in the previous section B was 10 ft.-L, C 1%, and a

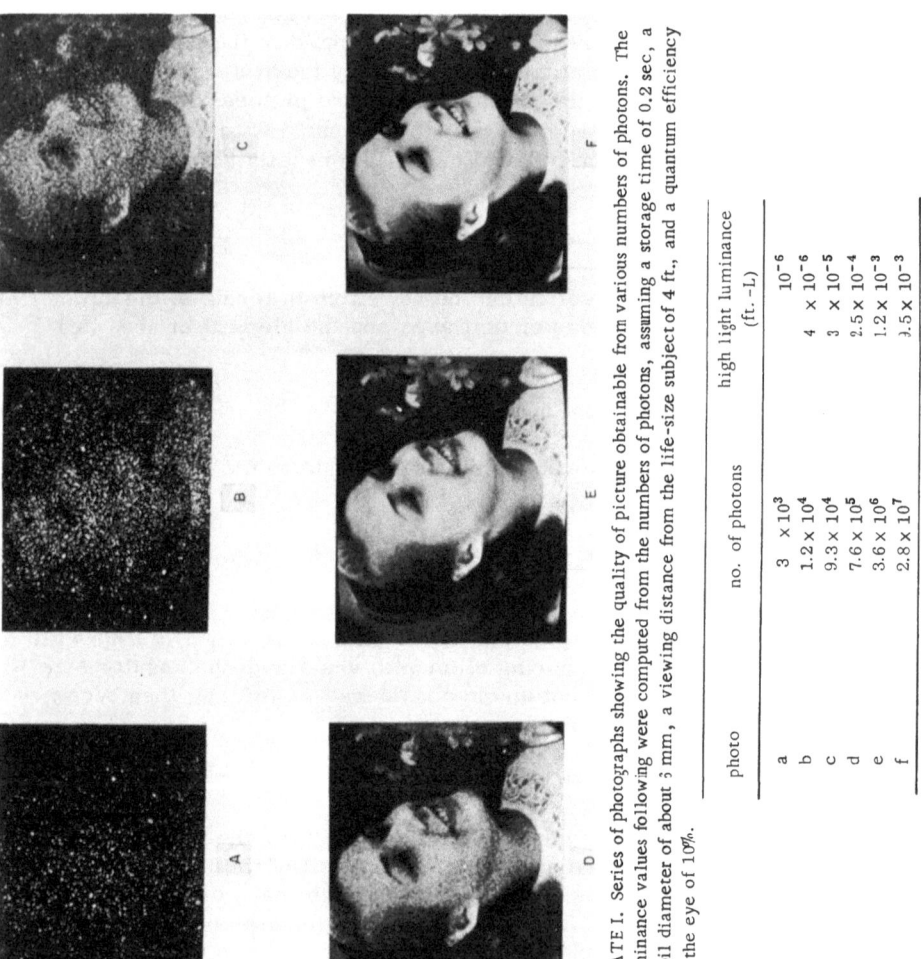

PLATE I. Series of photographs showing the quality of picture obtainable from various numbers of photons. The luminance values following were computed from the numbers of photons, assuming a storage time of 0.2 sec., a pupil diameter of about 3 mm., a viewing distance from the life-size subject of 4 ft., and a quantum efficiency for the eye of 10%.

photo	no. of photons	high light luminance (ft. –L)
a	3×10^3	10^{-6}
b	1.2×10^4	4×10^{-6}
c	9.3×10^4	3×10^{-5}
d	7.6×10^5	2.5×10^{-4}
e	3.6×10^6	1.2×10^{-3}
f	2.8×10^7	3.5×10^{-3}

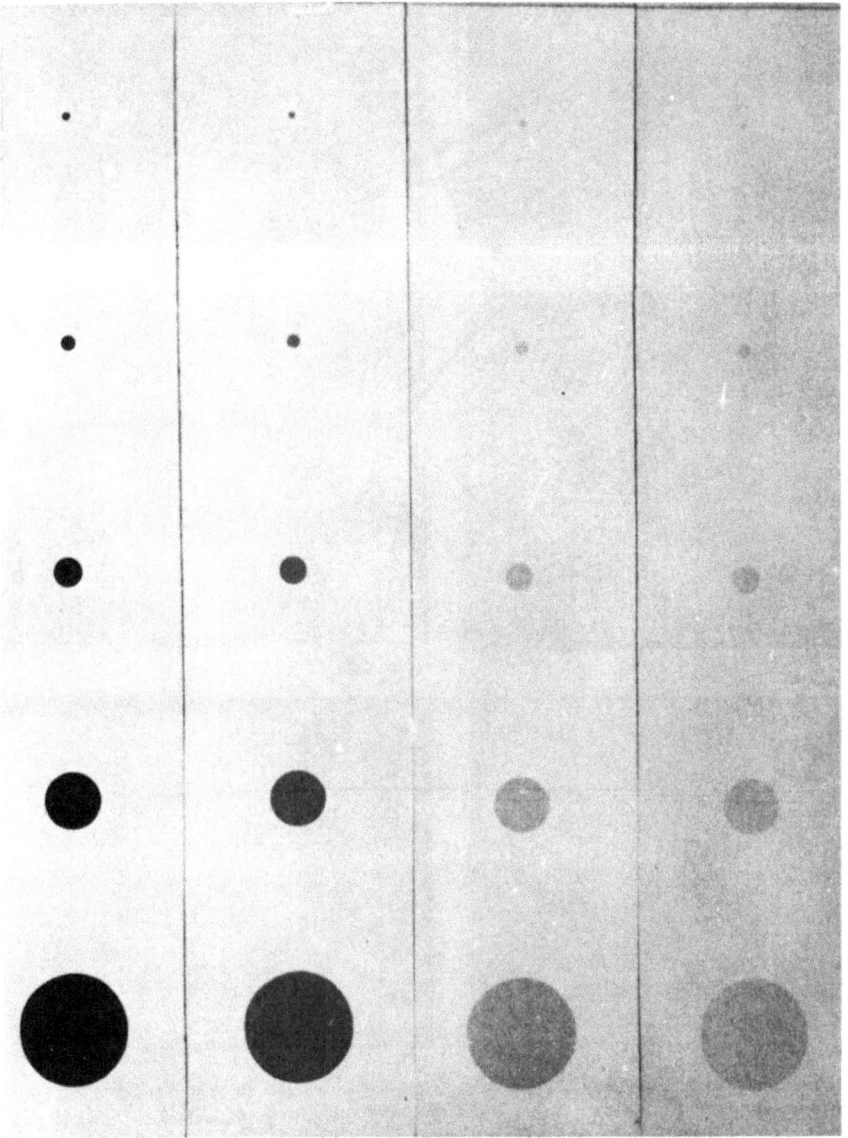

PLATE II. Test pattern used for measuring performance in series of photographs in Plate III.

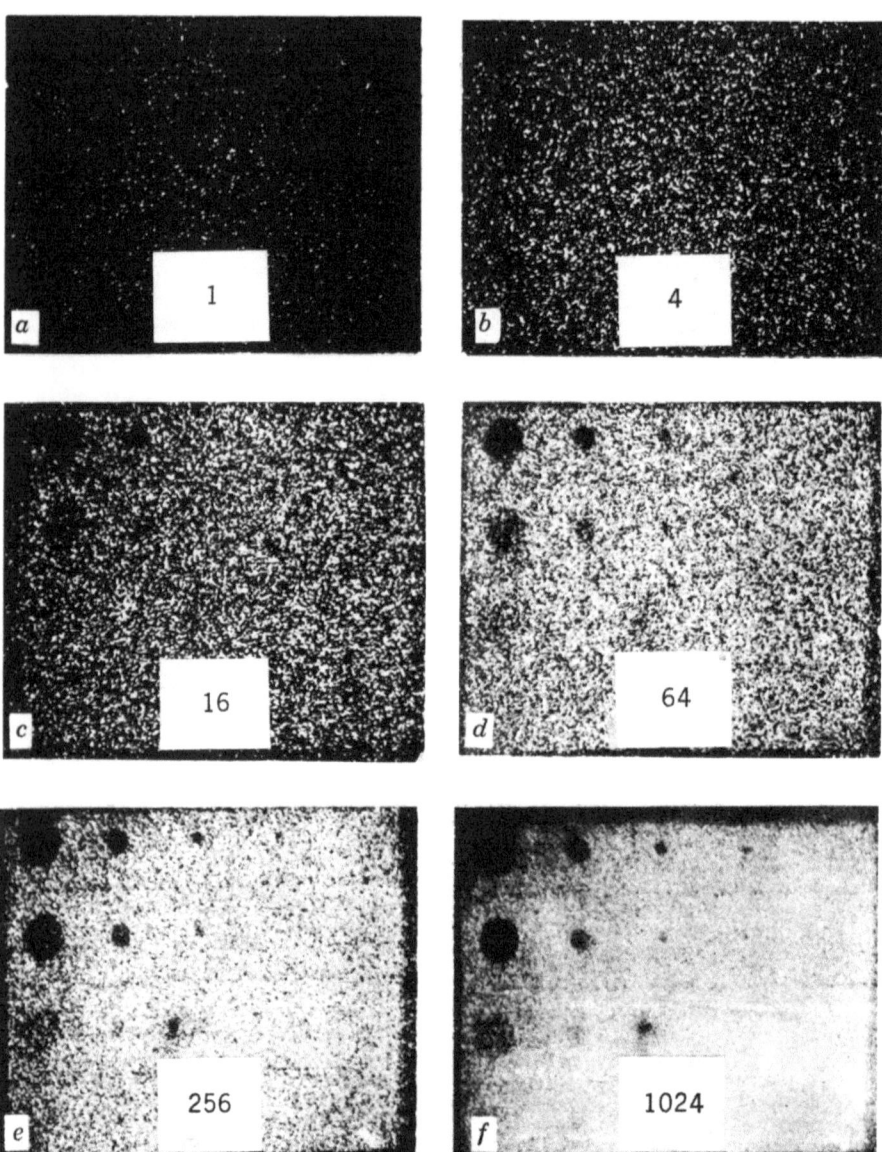

PLATE III. A series of light spot scanner pictures of the test pattern in Plate II. The pictures were photographed from the kinescope using camera exposure times starting from a of 1/16, 1/4, 1, 4, 16 and 64 secs respectively. The relative numbers of photons are indicated on each photograph.

2 min of arc. The value of the constant, then, for full use of every incident photon is 10. This value is based on a pupil diameter of 4 mm.

The series of photographs in Plate III were used to test Eq. (4) as well as to derive an experimental value for the threshold signal-to-noise ratio k. Plate III shows what a perfect eye would see when looking at the original pattern shown in Plate II under various illuminations. The adjective "perfect" refers to the fact that each white dot (visible particularly in the first three pictures of the series) is an actual trace of a single photon reflected from the original pattern and entering the photomultiplier. * The gain of the photomultiplier and subsequent television amplifier was sufficient to show traces of the individual photons on the kinescope from which Plate III was photographed.

The original test pattern, Plate II, consisted of black disks along the top row decreasing in diameter by a factor of two in each step. The second row is a repetition of the first except that the disks are now gray, with 50% contrast. In the third row the contrast is 25%, and in the fourth 12%. The pattern was so designed that at each level of illumination the demarcation line between the resolvable and unresolvable parts of the pattern should be a 45° line if Eq. (4) is valid. This demarcation line should also move one step to the right for each factor of four increase in illumination. The illumination in successive pictures was increased by a factor of four each time. Both these expectations were borne out in a careful examination of the originals of Plate III. Actually each picture of Plate III is in itself a confirmation of Eq. (4) and serves to give several independent estimates of the threshold signal-to-noise ratio. The slight shading in the pictures and the variations in the sizes of the white dots reflect experimental difficulties not entirely overcome but not having a serious effect on the conclusions.

4. Threshold Signal-to-Noise Ratio

The statement is frequently made in communications that a just detectable signal is one that is equal to the noise* of the receiving system: in brief that the threshold signal-to-noise ratio is unity. This is a quite reasonable assumption to make in the absence of experimental data. It is also a statement that might be erroneously "confirmed" if the noise is not properly defined. For example, the noise is sometimes loosely used to refer to the envelope noise on a cathode ray tube display. The so-called envelope noise is some six to eight times (depending on viewer and viewing conditions) larger than the accurately definable rms noise. Also, one may easily overlook certain integration effects that reduce the actual value of noise below that read on a meter. It is, for example, common to specify the signal-to-noise ratio in a television system by measurements on or by analysis of the wide-band amplifier. It must be borne in mind that this signal-to-noise ratio applies only to areas of picture element size viewed

---

* The television light spot scanner arrangement used to obtain the photographs of Plate III is described in [23]. In communications, the term "noise" is used to refer to the fluctuations in signal currents both in video as well as audio systems. "Noise" is used in the present paper interchangeably with "fluctuations."

in single pictures. In viewing an actual picture, the eye can integrate several successive television pictures and thereby improve the signal-to-noise ratio. Further, if elements larger than picture element size (as determined by the passband of the amplifier) are under consideration, the signal-to-noise ratio is higher in accordance with Eq. (4).

What can be stated here with confidence based on simple inspection of Plate III, is that a signal just equal to the rms noise is not detectable. If it were detectable it would mean that one could remove one of the photon specks in Plate IIIa or Plate IIIb and the removal of that photon would be detectable by another person. One photon in these pictures represents a signal of unity and also a noise of unity, since the noise is the square root of the signal. In fact, the questions of what the threshold signal-to-noise ratio means and what its approximate value is can also be answered by simple inspection of these figures.

One poses the problem: "What is the smallest number of photons that must be removed so that the removal is detectable?" The removal of photons from a given area corresponds to the insertion of a black disk of that same area in the original test pattern. By such an inspection of Plate III or by, what is the equivalent, noting the size of black disks that are just detectable in these photographs, it was concluded that about 25 photons had to be removed to get a clear indication that something was missing. The 25 photons represent a signal of 25 and a signal-to-noise ratio of 5. Hence, the origin of the value 5 for threshold signal-to-noise ratio. Another way of stating the problem is that there are already present in the uniformly illuminated areas of Plate III statistically formed " holes" in the distribution of photons of a magnitude approaching 25 photons. To avoid confusion with these spurious "black" areas a real black area must be somewhat larger.

These last statements can be made more quantitative. Consider a uniformly illuminated area and within this area consider a test area containing an average of 100 photons. The rms fluctuation associated with the test area is 10 photons. One now asks, "How many photons must be removed from the test area in order that the removal be detectable?" If one assumes that a signal-to-noise ratio of unity is detectable, only 10 photons need be removed. But then, one asks, "What are the chances of seeing another area, the same size as the test area, that has 10 photons missing by accident?" The answer is that approximately one out of three such areas picked at random will show this deviation. If the test area occupies a small fraction, say 10^{-5}, of the field of view, one will see about 3×10^4 spurious deviations. If one asks for a signal-to-noise ratio of 2, there will be about 10^4 spurious test areas. At a signal-to-noise ratio of 3, the number of spurious test areas is 300, and at a signal-to-noise ratio of 4 it is 6. Finally, a signal-to-noise ratio of 5 would reduce the chance of finding a single spurious test area in the whole field of view to less than one-tenth. Thus, threshold signal-to-noise ratios between 4 and 5 would appear to be safe. Moreover the threshold signal-to-noise ratio ought to become somewhat smaller, the larger the test area. The improved performance shown in Figure 1 for larger area test elements could be interpreted as reflecting this decrease in threshold signal-to-noise ratio. *

---

* Careful measurements by Tol [32] on a 2 mm test element led them to estimate the threshold signal-to-noise ratio at a value between 3 and 4.

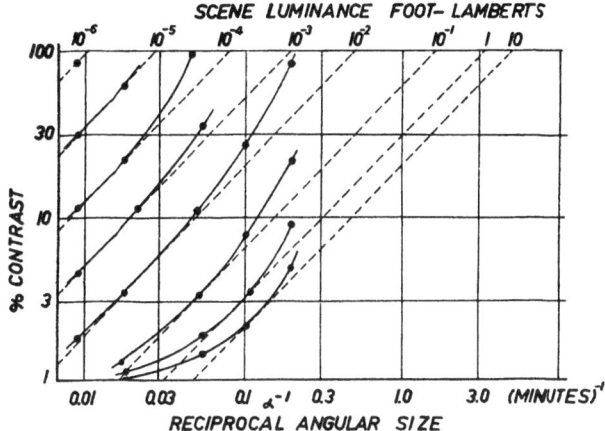

FIGURE 1. Comparison of performance data for the eye (computed from Blackwell) with ideal performance curves.

III. QUANTUM EFFICIENCY OF THE EYE

1. *Comparison with Visual Data*

Some of the most complete data on the performance of the eye have been reported by Blackwell [2]. Part of his data has been replotted in Figure 1 in a form to be readily compared with the fundamental relation (Eq. (4)) derived from purely statistical reasoning. The experimental data show the threshold contrast detectable as a function of the reciprocal angle subtended by the test object and at various scene brightnesses ranging from 10^{-6} to 10 ft.-L. The data were taken with white light and are uncorrected for pupil opening. Tangential to the best performance at each value of scene luminance are drawn the dotted lines with 45° slope representing the type of ideal performance expected from Eq. (4). The experimental curves depart from the 45° lines at both small contrasts and small angular sizes. Since the smallest detectable contrast and smallest detectable angle are set by local limitations in the eye and not by the statistical fluctuations in the incoming photons, one would expect the experimental data to depart from the statistically derived 45° lines as these limits are approached. Even at lower illuminations, where the statistically limited resolution is well below the limit set by the rods and cones, the eye must perform a quite ingenious technical feat in order to approach ideal performance. That is, for low-contrast areas it must pool the information from hundreds or thousands of receptors in order to perceive the low-contrast area, while at the same time it must reduce the pooling effect for high-contrast areas where higher resolution is statistically possible. It is conceivable that some of the incoming photons are received not in picture-forming receptors but in monitoring receptors whose function is to control the amount of local pooling.

From the 45° lines in Figure 1 and from data published by Reeves [21] on the size of pupil opening as a function of scene illumination, one can compute with the aid of Eq. (4) the effective quantum efficiency of the eye.

The meaning of an effective quantum efficiency of 10%, for example, is that the performance of the eye is equal to that of an ideal fluctuation-limited device making use of 10% of the incoming photons. The effective quantum efficiencies, rounded out to the nearest half per cent, are listed in Table I. They are shown for white light as derived directly from Blackwell's data in Figure 1. A second column of efficiencies is shown for blue-green light. ** These are simply the white-light efficiencies multiplied by three since blue light (or green depending on the luminance range) requires only one-third as many photons to produce a given luminance as does white light. To obtain the efficiencies in Table I, an exposure time of 0.2 sec and a threshold signal-to-noise ratio of 5 was used. If there is any change in these parameters as a function of luminance, the exposure time is almost certain to decrease and the threshold signal-to-noise ratio to increase with increasing luminance. Both these changes would be in the direction of reducing the change in quantum efficiency from low lights to high lights.

Two facts are of striking importance in Table I. First, the quantum efficiencies are relatively high, particularly at low lights. Second, there is remarkably little change in the quantum efficiency from 10^{-6} to 10 ft.-L. The factor of 4 change shown in Table I is an upper limit and is insufficient, as will be discussed later, to support the bleaching of the visual purple model for dark adaptation.

TABLE I. The Effective Quantum Efficiencies

Brightness ft.-L	Effective quantum efficiencies in per cent		
	White light*	Blue-green light 5100 A to 5500 A	Blue-green light at the retina
10^{-6}	3	9	18
10^{-5}	4	12	24
10^{-4}	3	9	18
10^{-3}	2	6	12
10^{-2}	2	6	12
10^{-1}	2	6	12
1	1.5	4.5	9
10	1	3	6

* As derived from Blackwell's data in Figure 1.

The last column of Table I is labeled "Effective quantum efficiencies for blue-green light at the retina." The values here are twice those in the previous "Blue-green" column. Since it is known that only about half the light incident on the eye arrives at the receptors, it is worth indicating this factor separately. The last column, then, is a measure of how much

** At high lights the maximum efficiency of the eye is in the green at 5550 A; at low lights it moves to the blue at 5100 A.

of a discrepancy must still be accounted for between the actual and the theoretically perfect performance of the retina as a picture device. A small part of this discrepancy can undoubtedly be ascribed to incomplete absorption of the light in the receptor elements,* to some reflection, and to some absorption by non-sensitive tissues separating the receptor elements. Perhaps a more significant source of the discrepancy lies in the many functions the eye performs. Color vision at scene luminances above 10^{-1} ft.-L is one of the most important.

There are other functions that might tap off some of the light onto monitoring cells. The adjustment of the high-gain process that must intervene between the receptors and the optic nerve pulses is one such function. Electrical recording of pulses in single nerve fibers by Hartline [10] has shown that some receptors are active only transiently when the light is increased or decreased but not when it remains constant. These elements might indeed be monitoring the magnitude of the high gain process. The additional function of monitoring the number of receptors that must pool their information has already been mentioned.

2. *Quantum Efficiency Derived from Absolute Threshold*

It has been known for a long time that only about 100 photons need be incident on the eye within 0.001 sec in order to give a visual sensation. To carry out the above experiment the eye is completely dark-adapted and then exposed to a series of light flashes. The intensities of the light flashes are at first below the threshold of seeing and later increased until the observer gives reliable reports that he sees the flashes. This measurement, then, is a measurement at absolute threshold — that is, a single white area against a black background.

It occurred to Hecht [11], and apparently independently to others [3, 33] that one could make use of the transition curve from "not-seeing" to "seeing" in order to compute the actual number of photons taking part in the visual sensation. Qualitatively, if only 1 or 2 of the 100 incident photons were finally used, the transition from not-seeing to seeing would be relatively slow. It would occupy perhaps a factor of 2 in incident light intensity. This argument is based on the statistical fluctuation to be associated with a small number of random events. At the other extreme, if all 100 of the incident photons were used, the transition would be relatively steep, occupying only about a 10% change in light intensity. The argument is quite reasonable. The results obtained, however, by Hecht [12], by Brumberg [3], and more recently by van der Velden [33] show a disturbing spread. Hecht concluded that only 7 out of the 100 quanta were used, Brumberg's value ranged from 25 to 50, while van der Velden has concluded that the number should be 2. Van der Velden has gone into

---

* If one takes the estimates of Hecht [12] or of Dartnall and Goodeve [6] for the light absorption by receptor elements (about 20%), the differences between the efficiencies in the last column of Table I and 100% are mostly accounted for (and in one instance, 10^{-5} ft-L are over accounted). The reader may find it difficult to believe, however, that the eye wastes 80% of the light.

great detail to discuss the mathematical statistics of this method. These arguments, the writer has not evaluated.* What does occur to the writer is a parallel between the problem of determining the quantum efficiency of the eye and the same problem for a photomultiplier.

One can, of course, easily measure the quantum efficiency of a photomultiplier by taking the ratio of the electron current leaving the photocathode to the photon current incident on the cathode. The parallel operation cannot readily be carried out for the eye. There is, however, a second simple method for measuring the quantum efficiency of the photomultiplier. One can measure the signal current (i. e., the average current coming out of the multiplier) and compare this with the noise current coming out of the multiplier. This ratio is the signal-to-noise ratio in the output stage of the photomultiplier. It should also be closely equal to the signal-to-noise ratio of the current leaving the photocathode of the photomultiplier. Knowing the signal-to-noise ratio of the cathode current, the current itself may be directly computed. The ratio of the cathode current to the current of photons incident on the cathode then gives the quantum efficiency of the photocathode. This procedure is essentially what the writer has carried out for the eye in order to compute the quantum efficiencies shown in Table I. That is, the signal-to-noise ratio of the output of the eye was estimated from the half-tone discrimination of the eye; from this ratio the number of photons effective at the retina was computed and compared with the number of incident photons to get a value for the quantum efficiency.

What is to be emphasized about this procedure is that it can be carried out at scene brightnesses well above absolute threshold. At these scene brightnesses, the noise from the incoming photons is likely to be greater than possible local noise sources in the retina. The latter can be significant and can mask the measurement of quantum efficiency at extremely low values of scene brightness.

To continue the parallel with the photomultiplier, if one tried to carry out the method of Hecht and others to determine its quantum efficiency, one would use small flashes of light on the photomultiplier until a reliable signal was obtained. This procedure would not necessarily give the quantum efficiency of the photomultiplier. In the absence of light, one can still observe currents in the output stage of the photomultiplier. These signals come from thermionic emission from the photocathode, from leakage currents, from stray gas ions, and possibly from cosmic rays. The magnitudes of these local noise sources would normally confuse any measurement of the quantum efficiency of the photomultiplier

---

* Baumgardt [1] has repeated and confirmed van der Velden's measurements. He also confirms the conclusion that the basic retinal units (a unit may be a collection of hundreds of receptors) is triggered off by two photons. He goes on to show, however, that it would not alter the agreement with experiment to assume that two or even more of these basic units need to be triggered off before a recognizable sensation is transmitted to the brain. This argument of Baumgardt's would allow van der Velden's results to be consistent with the higher quantum efficiencies shown in Table I. If only one basic unit needed to be triggered, the best quantum efficiency that van der Velden could offer would be perhaps 2%, to be compared with the 12% in the blue-green column of Table I.

carried out at absolute threshold. In brief, the local noise sources are greater than the noise of the incoming photons. At light intensities well above absolute threshold, the reverse is true. The writer, for these reasons, is inclined to place more reliance on quantum efficiency measurements of the eye carried out at light intensities well above absolute threshold than on those carried out at absolute threshold.

IV. SOME PROBLEMS OF VISION

1. *Primary and Secondary Processes*

It is essential to an understanding of the eye that one distinguish between primary and secondary processes. This is just as true for an understanding of its intimate chemistry as it is for an understanding of its performance. Separation of primary and secondary processes is a well recognized prelude to an analysis of those man-made competitors to the eye — photomultipliers, photographic systems, and television camera tubes. All of these devices, as well as the eye, are subject to the same statistical limitations of the incoming stream of photons and to the same principles of physics and chemistry in the subsequent development of detectable signals.

The primary process is clearly the useful absorption of incident photons. Present knowledge of quantum physics makes it almost certain that the photons are absorbed by electrons rather than ions. The electron may be liberated into the vacuum as in a photoemitter; it may be liberated within the solid as in photoconductors and as in silver bromide crystals of photographic film; it may be raised to an excited state from which its energy may cause the dissociation of molecules, or it may trigger off one of many possible photochemical reactions. It is not important which of these types of absorption takes place in the primary process. What is important is that all of the photons be absorbed in the same way and in such a way as to be countable by their subsequent physical or chemical effects. Only then can all of the information be drained out of the incoming stream of photons.

In the photocathode of a photomultiplier, only a fraction of the absorbed photons succeed in liberating electrons into the vacuum where they can be counted by the multiplier. In photographic film, only a fraction of the internally liberated electrons arrives at opportune spots in the crystal where they can contribute to (or be counted by) the developing of grains; in a photoconductor, some of the liberated electrons may be associated with longer-lived, higher-gain centers than others, thereby pre-empting the counting process. Consequently it is not sufficient that all of the photons be absorbed by the same type of electronic process; they must also be able to contribute equally to the subsequent train of events. The fraction of incident photons so absorbed is the quantum yield of the primary process and determines the maximum fraction of incoming information that can finally appear in the output of the picture recording device. The intermediate or secondary processes may deteriorate the information but they cannot recover the information lost in the primary process.

The secondary processes are those that intervene between the primary absorption of photons and the final presentation of the picture. The energies of photons are microscopic compared with the energies normally associated with perception. On the microscopic scale, the energy of a photon is large compared with the thermal energy of an atom and can thereby trigger off normally improbable* events to make itself visible on a macroscopic scale. A well known example is that of the silver bromide grains in ordinary photography. The energy of the few photons absorbed by the silver bromide grain is sufficient to change the state of only a few atoms of silver. To become visible, the state of all 10^{10} atoms of silver in the grain must be altered. Thus the secondary process here provides a gain or amplification of ten billion times. The same order of gain intervenes between the photosurface of a television camera tube and the final viewing audience.

A very rough approximation of the energy of a single pulse in the optic nerve can be made from the following estimated values: pulse height 10^{-3} volts; nerve fiber resistance 10^3 ohms; and pulse duration 10^{-3} sec. This estimate yields an energy of 10^{-12} joules or a value over a million times larger than that of the few photons known to be able to trigger the pulse. Undoubtedly, then, the secondary processes in the eye must provide a considerable gain in energy just as in the other systems of picture recording.

Not only is it common in picture recording systems for the secondary processes to provide a high gain, but this gain must be variable. The energy level of presentation as measured by the brightness of motion picture or television screens is relatively constant, while the energy level in the primary process as determined by the scene or image brightness may vary over many powers of 10. The variable gain of the secondary processes must match this variable ratio between final presentation and initial image brightnesses.** (In photography and television, the gain of the secondary processes includes also the ratio of final picture area to initial image area.)

In photography and television the range of scene brightnesses transmitted is of the order of a thousand, i.e., from about 1 to 1000 ft.-L. In the case of human vision, this range is on the order of a billion, extending from 10^{-6} to 10^3 ft.-L. The importance of a variable gain mechanism in the secondary process of the eye is accordingly even greater than in photography or television where it is already known to exist. In spite of our lack of knowledge about the presentation level in the brain, we can be reasonably confident that it does not vary by the factor of a billion just cited. In the absence of the telltale objects of common experience, our ability to discriminate absolute brightnesses in a uniformly lit white enclosure is notoriously poor. Furthermore, the

---

* The probability that in one second an atom takes on from thermal vibrations the energy of a photon of green light is less than 10^{-10}. This means that gains approaching 10^{20} would be stable against thermal triggering.

** If the quantum yield of the primary process is less than unity, the same fractional value of image brightness is to be used.

measurements of Hartline [10] on the optic nerve fibers of the horseshoe crab show that the electrical signals passed on to the brain vary only as the logarithm of the light intensity.

The arguments for the existence of a variable gain mechanism between retinal receptors and optic nerve fibers may appear to have been labored too much. The writer can only say that such a mechanism must play an important role in many visual phenomena (two of which are discussed) and he has rarely seen the variable gain mechanism considered in other analyses of vision. *

2. Dark Adaptation

Once the separation of primary and secondary processes is recognized, the way is opened for a simple interpretation of some well known visual phenomena. Dark adaptation is one example. Its characteristics have been repeatedly explored. The major facts are these: if a person enters a dimly lit room after having been exposed to normal high-brightness levels of about 100 ft.-L, he finds at first that he sees little or nothing. After a time he begins to discern objects not at first visible. After a half hour he can see objects whose brightness is perhaps a thousand times lower than what he could see when he first entered the darkened room. The improvement in vision with time in the dark is called "dark adaptation." The problem is how to interpret the factor of a thousand or more improvement in seeing.

Hecht [11] and others have reversed the problem by suggesting that one starts with maximum visual sensitivity in the dark. In the dark or near absolute threshold the sensitive material in the retina, called visual purple, absorbs about 20% of the incident light. As the ambient brightness is raised, the visual purple tends to bleach out and accordingly to absorb less of the incident light. **

Finally at high light intensities, it absorbs less than one thousandth of what it absorbed near absolute threshold. In brief, the true sensitivity of the eye has been depressed by a factor of 1000. If the process is now reversed and one enters a darkened room, time is required for the visual purple to be regenerated. Dark adaptation is then identified with the regeneration of visual purple.

The significance of the "bleaching of the visual purple" model for dark adaptation is far reaching. It means that the sensitivity of the eye at normal seeing levels is one-thousandth what it is at low light levels. If one takes the quantum efficiency of the eye at low light levels to be about 10%, then the quantum efficiency at high light levels must be less than 0.01%. One compares this value with values in the neighborhood of a few per cent shown in Table I. The latter were deduced from purely statistical considerations without reference to any particular assumptions about the retinal mechanism. The values in Table I are absolute minimum values

---

* De Vries [9] makes passing mention of possible triggering action of the absorbed photons. See also Schade [27] for a thorough comparative analysis of visual and television systems. See also Talbot [30].

** For a recent review of the visual purple mechanism see Wald [34].

for quantum efficiencies. The eye must absorb at least these fractions of the incoming light in order to see what it does. (It may actually absorb more light to be used for other visual functions.) Whatever uncertainties are involved in computing the quantum efficiencies in Table I, it is highly unlikely that the error would be even a significant fraction of the factor of several hundred needed for reconciliation with the bleaching theory.

To restate the argument in a positive manner, the change in quantum efficiencies in Table I from 10^{-6} to 10 ft.-L is only a factor of 3. A factor of more than 1000 is needed to account for the known facts of dark adaptation. Changes in the quantum efficiency or sensitivity of the primary process cannot, therefore, account for dark adaptation (see also Rushton [25]).

If dark adaptation is not assignable to the primary process, its interpretation must be found in a secondary process. One such secondary process has already been described that appears to fit readily, i.e., the variable gain mechanism that amplifies the energy of a few absorbed photons to the energy level of optic nerve pulses. At low lights, this gain must obviously be set high. As the light intensity is increased, the gain is reduced as evidenced by the measurements on the energy flow in the optic nerves. The frequency of nerve pulses increases slowly, i.e., logarithmically with increasing light intensity [10].

The reduction in gain does not mean a lower sensitivity but only that less amplification is needed to present the information to the brain. A common parallel is found when we tune a radio receiver from a weak to a strong station; the gain of the receiver is reduced to match the presentation level of our hearing. The reduction in gain does not reduce the information transmitted. If we now tune back again to the weak station, we find that it is not audible until the gain is once more increased. Similarly, if we re-enter a darkened room after being exposed to high light intensities, time is required to reset the gain mechanism at its maximum value. The steady improvement in seeing reflects the steady increase in amplification provided by the gain mechanism. Whether one describes this transient as an increase in sensitivity (a description the writer regards as confusing) or as a temporary mismatch of gain is not so important as it is to recognize that it is transient, that it is associated with a secondary process, and that the steady-state sensitivity of the eye changes very little from low lights to high lights.

It is worth likening the gain mechanism to one other familiar system. The resolving power of a microscope is commonly defined in terms of its objective lens. The magnification that results from the objective is to some extent incidental. A minimum value, called the useful magnification, is needed to present to the observer all of the information resolved by the objective lens. Magnifications greater than this value may add convenience but do not add information.

There is no difficulty in conceiving of a high gain mechanism in the retina. The chemical literature provides numerous examples of catalytic processes in which a small number of atoms of one type can facilitate reactions in large numbers of atoms of another type. Moreover, the human system is already skilled at devising the powerful triggering mechanisms that allow heavy muscular activity to be controlled by the relatively micro-scopic energy of nerve impulses. Finally, the decrease in gain at higher light levels would follow naturally if the amount of material available to be triggered is limited. The writer wonders whether the substance called visual purple is not the triggered material responsible for the high gain rather than the material responsible for the absorption of light. Bleaching or exhaustion of the visual purple at high light intensities would then be more easily understandable.

The present model for dark adaptation has separated the visual processes into a primary process (the absorption of photons) that controls the amount of information received by the eye and a secondary process, a gain mechanism, that determines the level of presentation of the information. These are separate operations and conceivably can act independently. In fact, it is the time lag in resetting the gain mechanism that is used to account for dark adaptation. Another example of independent action is discussed in the following section.

3. *Subjective Brightness and Information*

It is known that at high brightness, the eye sees closely the same fine detail in red, green, or blue light as in white light. Luckiesch and Taylor [15] performed the following test. They started with equally high bright-nesses of red and blue light and observed that the fine detail visible under both illuminants was the same. Next, they inserted neutral filters in front of each illuminant to reduce its radiance by a factor of 100. The detail visible under each illuminant was reduced but remained mutually the same. On the other hand, the apparent brightness of the blue illuminant was considerably greater than the red illuminant. The quantitative results in this test are not of particular concern here. What is significant is that qualitatively the information transmitted by the red and blue illuminants did not change at the same rates as did their apparent brightnesses.

A simple interpretation of the above test can be made in terms of the primary and secondary visual processes. The insertion of a neutral filter reduced the number of photons reaching the eye from each illuminant by the same factor. Hence, the information transmitted by the two illumi-nants (determined by the number of photons absorbed in the primary visual process) remained equal. The brightness, on the other hand, is determined both by the number of photons absorbed and by the gain of the secondary process. If the secondary processes for the two illuminants are taken to be different and to provide different gains at low brightnesses, the higher brightness of the blue illuminant can be accounted for. For example, it is known that low light vision is provided by the system of rods as opposed to the cones. The rods must accordingly have a higher gain secondary process associated with them than do the cones. Further, the rods are sensitive to blue and green light but have very little response for red light.

4. Seeing Noise

It is natural to inquire about the magnitude of the gain process in the eye. If the performance of the eye is to be limited by the noise of the incoming photons, the gain must at least be sufficient to make this noise visible. A smaller gain would entail a loss of information. A larger gain, on the other hand, would add no more information but might contribute the annoyance of a prominent noisy background. The series of photographs in Plates I and III, particularly the first three in each series, show the kind of pictures we would see at low lights if the gain in our visual system were considerably higher than it is. If the gain were freely at our disposal, it is possible that some individuals might prefer such a gain setting. Most people, however, would set the gain carefully at the point where the noise was just visible. At least this would be true if we can take television viewing practices as a guide. Here, the gain is freely at the disposal of the viewer and one seldom finds it set so high that the background noise becomes distracting.

Whether it was through the influence of a kind of personal choice or a result of natural economy, the evolutionary processes appear to have set the gain at each brightness close to the threshold of noise visibility. At normal brightness levels, it would be difficult to persuade oneself that visual noise can be perceived. At low levels, however, in the neighborhood of 10^{-4} ft.-L, the writer is confident from his own observations that visual noise is readily "seeable." This is particularly true for large uniformly lighted areas. A dimly lit wall takes on a fluctuating granular appearance reminiscent of a noisy television picture or a grainy motion picture screen. An important corollary to this observation is that the noise is absent in neighboring black areas. Whatever one may conclude about the fluctuations, at least they do not originate as independent local noise in this observer's visual system.

The writer knows of only one other published interpretation of the graininess of large dimly lit areas in terms of visual noise. The terms used by de Vries [8] to describe the granular appearance are quite similar to those used here. (De Vries also published one of the first attempts to account for visual performance at high lights in terms of photon noise. His quantum efficiencies, based on a signal-to-noise ratio of unity, are almost a hundred times smaller than the values shown in Table I.) The writer has also found from an informal canvass of his colleagues that most of them describe the appearance of dimly lit scenes in terms consistent with a visual noise interpretation. The advantage of large uniformly lit areas for seeing noise is well known from experience with television and motion picture screens. In fact, a device used to obscure the noisiness of early motion picture films was to break the background scene into a high-contrast, checkerboard array of wallpaper patterns or objects. The discrimination of the eye for halftones is markedly deteriorated in the neighborhood of a sharp black to white transition.

5. Eigenlicht

One frequently finds the statement that a completely dark room does not appear as dark as one in which there are local patches of light. In fact, in utter darkness, the sensation of a gray light, Eigenlicht (self-light of the retina), is often reported. Another sensation in utter darkness is that of visual strain.

All of these observations are consistent with what one might expect from a high gain mechanism in the eye. In a partially lit darkroom, the magnitude of gain is set by the visible areas. The other areas then appear completely lightless. If the room is devoid of any visible light, there is nothing objective by which the eye can calibrate its gain mechanism. Under these conditions, one can readily believe that the gain is turned up to its maximum useful value until local system noise or disturbances are visible. This would account both for the appearance of Eigenlicht and for the feeling of strain in trying to distinguish objective from subjective sources of light sensations. Indeed, knowing the tendency of observers to see regular patterns in an array of purely random noise, one can speculate that the source of some optical illusions may lie in the random character of the Eigenlicht.

6. Visibility of Noise in Television and Motion Pictures

There is no doubt that the conditions under which we see noise in a television picture (or its counterpart, graininess, in a motion picture) are of considerable economic consequence. The size and brightness of picture screens, the grain size of photographic film, the intensity of studio lighting, and the transmitter power of a broadcast station all hinge on the visibility of noise. The aim, of course, is not to see the noise.

The most succinct condition for not seeing noise in a picture is that the noise originating in the outside picture be less than the noise of the photons entering our eye. It is often more convenient to deal with the parameter signal-to-noise ratio, i.e., the ratio of average brightness to statistical fluctuations in brightness. In these terms, we ask that the signal-to-noise ratio of the television picture or motion picture be greater than the signal-to-noise ratio of our visual image.

Some immediate consequences of this condition for seeing noise are familiar experiences to most people. Because they are usually brighter than motion pictures, television pictures must have a better signal-to-noise ratio to avoid being considered noisy. Similarly, as the brightness of motion picture screens has been increased, finer grained (less noisy) films have had to be used. The converse experience is an interesting one. If one is presented with a noisy television picture, he can convert it into a "good," noise-free picture by the simple expedient of holding a neutral filter in front of his eyes [22]. It is important to have the neutral filter covering the observer's eyes rather than just the television screen. If the neutral filter covers only the television screen, it is still true that the noise is "filtered" out, but then the observer concludes that the picture is poor because it is dim. The judgment "dim" is made by comparing it with surrounding objects. If, on the other hand, the neutral filter covers the observer's eyes, the brightnesses of both the television screen and the surroundings are reduced by the same factor. The television picture does not suffer by comparison with the surroundings. Further, in so far as one's judgment of the absolute level of brightness is poor, the first order effect of the neutral filter is not that of reducing brightness but only of removing the noise.

The action of a neutral filter on the visibility of noise is especially efficient when one compares it with other methods of noise reduction. For example, large reductions in noise power, as read on the meter or as observed on an oscilloscope, can be made by reducing the bandwidth of the television system. The reduction in bandwidth, of course, cuts down the fine detail in the picture as well as the fine grained noise. Similar large reductions in noise power can be conveniently effected by slight defocusing of the scanning beam in the television receiver or by slight optical defocusing of the motion picture projection lens. While the total noise power may be reduced by a factor of 10, the noisiness of the picture remains substantially unchanged. The reason is that bandwidth reduction and optical defocusing selectively remove the noise from the high-frequency end of the spectrum. That is, both the fine grained noise and the fine grained picture detail are filtered out leaving still intact the noise associated with larger areas. But, in a well balanced system, the eye sees noise in large areas as well as in small areas. Thus, cutting out the small area noise and picture detail does not alter the annoyance of noise in the larger areas. The observer concludes that in spite of the large reduction in total or integrated noise, the picture is still noisy.

In contrast to bandwidth reduction or optical defocusing, the use of a neutral filter reduces the visibility of noise uniformly for large as well as small areas. By Eq. (4), a reduction in brightness (B) reduces the halftone discrimination of the eye equally for all values of elementary picture area. The neutral filter has, incidentally, one other effect in its favor. The apparent contrast of the picture is increased, because the gamma (in the photographic sense) of the eye is larger at lower scene brightnesses.

7. Light Amplifiers

The phrase "light amplifier" has a certain magic sound that has encouraged extravagant speculations of being able to see in substantial darkness. "Given a device that intensifies the light just before it enters our eyes, what is to prevent us from extending our vision indefinitely?" The physician who spends almost half an hour dark-adapting his eyes and then is only barely able to discern the telltale patterns on a fluoroscopic screen must surely wonder how much he could profit from such a device. It is here that a recognition of the quantum limitations upon vision finds its most direct application. A light amplifier can provide a significant but not an unlimited gain in seeing.

The simplest estimate of the gain to be derived from a light amplifier comes from Table I. The quantum efficiencies shown in Table I for white light and for blue-green light are in the neighborhood of 5%. This means that the eye is already apprehending 5% of the total information incident on it. The most that a light amplifier interposed at the eye can do is to increase the 5% utilization of information to 100%. Sturm and Morgan [28] and Morton et al. [18] also concur in estimating the useful gain of a light amplifier to be about 20 times.

To realize a significant gain over the human eye, the photosensitive element of the light amplifier must have a correspondingly higher quantum efficiency than the eye. This is generally not true of the photoemissive cathodes used in vacuum type light amplifiers. Their quantum efficiencies for white light approach closely that of the eye. On the other hand, light-sensitive elements making use of photoconductivity can achieve quantum efficiencies close to 100% [35].

There are other conveniences in using a light amplifier that are not to be confused with the fundamental gain in information of a factor of 20. A light amplifier allows the physician to dispense with the time for dark adaptation. A light amplifier can make use of lenses having a larger diameter and light gathering power than the eye lens. It is true that "night glasses" do the same for the eye directly. However, it is more convenient to attach those heavy optics to a light amplifier. The light amplifier also permits the observer to be at a remote location, possibly safer and more comfortable.

Light amplifiers have a variety of forms. Electron image tubes [18] have received the most attention. More recently, several solid-state, flat-screen light amplifiers have been described [13, 20]. One must not overlook the fact that any photographic system is potentially a light amplifier. In so far as the final viewed picture is brighter than the original, light amplification has taken place. This is not usual for ordinary photography. However, one frequently finds television pictures displayed at ten or more times the brightness of the original scene. A common experience is that the waning minutes of a late-fall football game often take on a brighter aspect in the picture presented by a television receiver than in the original.

V. SOME PROBLEMS FOR FUTURE RESEARCH

The vast amount and the fruitful variety of research centered on the visual process needs no rehearsal here. It is perhaps sufficient to mention that in the last century contributions have come from biologists, chemists, physicists, biochemists, biophysicists, zoologists, and a variety of medical specialists. The proposals that follow are meant in no way to reflect on the significance of the body of knowledge already amassed or being acquired. What can, in all propriety, be suggested is that a sense of proportion would surely redistribute some of the effort toward a better understanding of the role of quantum limitations in vision. Such an understanding will not of itself uncover the visual mechanism. It can, however, by its very independence of mechanism and by its severe restrictions on the economy of photons help to eliminate those proposed mechanisms that are too wasteful of the photon stream.

1. Quantum Efficiency

The quantum efficiency of the eye is a measure of the minimum fraction of the incident light that the eye must use (independent of mechanism) to enable it to see what it does, in fact, see. Because this fraction is close to unity, it becomes a powerful tool in discriminating against those visual mechanisms that are too profligate of light energy.

Consider, for example, three commonly quoted factors: a factor of $\frac{1}{2}$ giving the fraction of incident light reaching the retina; a factor of $\frac{1}{3}$ if there are three different types of receptors for effecting color vision; a factor of about $\frac{1}{5}$ for the fraction of light absorbed by the light-sensitive material in the retina. If these three factors are combined, the maximum quantum efficiency of the eye could not exceed 3%. But the middle column of Table I based simply on what the eye sees shows quantum efficiencies greater than 3%.

One may, indeed, take issue with the accuracy of the results listed in Table I. And that is precisely the argument intended here. A quantity of such fundamental importance as the quantum efficiency of the eye merits more than one set of investigations by one investigator. This statement does not overlook the several papers [1, 3, 9, 11, 33] reporting on the minimum number of quanta used by the eye at absolute threshold. Table I, if anything, de-emphasizes measurements at absolute threshold, for it reports quantum efficiencies over a range of ten million in light intensity. The quantum efficiencies in Table I were computed by comparing what the eye sees, given a certain number of quanta, with what an ideal device would see with the same number of quanta.

There are other more direct ways of measuring quantum efficiencies. These involve a simple side-by-side comparison of the numan eye with other picture-recording devices of known quantum efficiency. For example, let an observer and a television camera view the same test pattern side-by-side. The observer need then only compare what he sees of the test pattern with what he sees reproduced on the television monitor to decide whether his quantum efficiency is greater or less than that of the television camera. The great virtue of this method is that one is not dependent on a knowledge of the storage time of the eye or of the threshold signal-to-noise ratio. Since the same observer compares the original and reproduced patterns, the same storage time and threshold signal-to-noise ratios are effective and their influence on the comparison cancels out.

It is true that there are some uncertainties about the quantum efficiency of the television camera. For this reason the supporting evidence from other comparisons is desirable. The increased sensitivity of recent photographic films would make a comparison between an observer and a motion picture camera of special interest. Finally, the comparison of an observer and a light spot scanner as described in [23] has the virtue that the quantum efficiency of the photomultiplier pickup can be accurately defined. What one would prefer in this arrangement is a scanner giving visible light in the range of 5000-5500 A. Since present cathode ray tube scanners emit in the near ultraviolet, a mechanical scanner would have some advantage.

2. *Threshold Signal-to Noise Ratio*

The significance of the threshold signal-to-noise ratio has already been described in an earlier section. It need only be mentioned that independent estimates by other observers would be helpful in arriving finally at a reliable figure. A quite extensive set of measurements was made by Tol et al. [32].

They have concluded that the threshold signal-to-noise ratio lies between 3 and 4 as opposed to the value 5 proposed by the present writer. It is likely that the threshold signal-to-noise ratio depends upon the angular field of view as well as upon the scene brightness. Little or nothing is known of these dependencies.

3. *Storage Time of the Eye*

There is reasonably good agreement that the physical storage time of the eye is close to 0.2 sec and that it varies little from extreme low lights to high lights. It is worth trying to confirm this value in other ways. There is an interesting method of making an independent check that, to the writer's knowledge, has not been carried out.

Consider a test pattern like that shown in Plate II and photographed on a single frame of grainy film. That is, only part of the test pattern is detectably reproduced owing to the film noise. If now the test pattern is photographed on successive frames by a motion picture camera and projected as a normal motion picture, the observer should see more of the test pattern. The increased visibility comes from integrating out the noise of successive frames and is a measure of how many frames are so integrated or stored. Tol et al. [32] present evidence that the physical storage time of the eye can be increased significantly by memory. An exaggerated example of integrating out the noise is the series of photographs in Plate III. Here, the first photograph in the series represents one field of a television picture. The improved reproductions in the later photographs of the series were obtained by letting the camera record or store an increasing number of frames simply by increasing its exposure time.

4. *Color Vision*

It is known that below one-tenth of a foot-lambert one's ability to distinguish colors rapidly disappears. It would be interesting to know how much of a role fluctuations in the incoming photon stream play in this transition from color to monochrome vision. Modern electronic color television systems would be an excellent guide, especially those color-slide scanners in which the light transmitted by the color slide is picked up by three photomultiplier tubes [29]. These would allow one to observe how rapidly the color reproduction is obscured by photon noise at low illuminations.

The transition from color to monochrome vision at low lights may also have arisen from a "tactical decision" on the part of the evolutionary processes. It is true for most man-made picture recording systems that more light is wasted by reproducing in color than by reproducing in monochrome. If the same is true for the eye, it could easily have been more important for survival to transmit at low lights a moderately well defined monochrome picture than a poorly defined color picture.

5. Variable Gain Mechanism

The arguments of the earlier sections have emphasized that the understanding of dark adaptation is likely to come not from the primary mechanism of light absorption in the retina but from the secondary reactions that are triggered off by the absorption of light. The absorption coefficient of the light-sensitive material is not likely to vary much from high lights to low lights. The magnitude of the triggered reactions must on the other hand undergo large changes.

6. Optic Nerve Pulses

Information is transmitted from the retina to the brain by a train of nerve pulses all having the same amplitude. The number of these pulses per second is a measure of the light intensity at the retina. But the measurements of Hartline [10] have shown that the frequency of transmission increases only slowly, possibly logarithmically, with increase in light intensity. A consequence of this relation is that at higher light intensities each nerve pulse represents a larger number of photons. Since the percentage fluctuations decrease as the average numbers of photons increase, the timing of each nerve pulse should become more precise. At the other extreme, in low lights where the nerve pulses represent one or a few photons the spacing of pulses should be quite random.

A careful examination of the trains of nerve pulses at various light levels should reflect the signal-to-noise ratio of the photon stream. At low lights, improvements in signal-to-noise ratio may be conveyed chiefly by an increase in the number of pulses per second. At high lights, improvements in signal-to-noise ratio may be conveyed chiefly by a greater regularity of spacing of the pulses. Measurements of the signal-to-noise ratio associated with the train of nerve pulses could yield directly the number of photons giving rise to these pulses. The electronic parallel is that a measurement of the signal-to-noise ratio of the output current of a photomultiplier gives directly the number of electrons per second leaving the photocathode or, what is the equivalent, the number of photons per second that are absorbed by the photocathode and emit electrons.

7. Visibility of High Energy Radiations

X-rays are observable by direct absorption in the retina. This is not particularly surprising. For example, there are many materials known to become photoconducting or luminescent when exposed to optical radiation. In all these materials, the same effects can be brought about by a variety of electron-exciting radiations including X-rays, gamma rays, alpha rays, beta rays, and other nuclear radiations. While optical radiations result in only one excited electron per absorbed photon, the high-energy radiations may excite hundreds or thousands of electrons for each high-energy photon or particle absorbed. Since the eye is known to respond at absolute threshold to only a few photons or electron excitations, it is surprising that no one has reported seeing any of the high-energy nuclear radiations directly. In particular, the frequency of occurrence of cosmic rays is such that one might expect to see an occasional track traced out in the retina by a cosmic ray striking almost parallel to the retina. Is it possible that evolutionary processes have developed some ingenious mechanism for discriminating against cosmic radiation since no useful purpose would be served by detecting it?

VI. VISION AND EVOLUTION

In a very real sense one can be confident that evolution has attained close to an absolute goal in the visual process. The absolute goal is to perceive all of the information contained in the stream of photons incident on the eye. The eye approaches this goal in the range of wavelengths of 5000 to 5500 Å. Not only the eyes of humans but also those of lower forms of animals have crowded close to this absolute limit.

It is true that many man-made devices have extended the range of human vision. These extensions include sensitivity to other spectral ranges (infrared and ultraviolet), longer storage times (photographic film), longer focal lengths (telescopes), shorter focal lengths (microscopes), shorter resolving times (photomultipliers), and larger light-gathering power (Schmidt lens). It is also true that other animals have a finer resolving power than the human (e.g. hawks) or a higher-speed eye lens (e.g. owls). In all of these instances, it is not so much a question of improving on the performance of the human eye as it is of choosing another theater of operation. The parameters: storage time, spectral range, and focal length define the field of application of the optical device. Once having made a selection of these parameters, the fundamental performance is measured by the ability of the optical device to make use of all incident photons. This the human eye does almost to perfection.

The particular choice of optical parameters for the human eye has been made for obvious evolutionary reasons. Its spectral response is peaked to match the sun's radiation in the daytime and shifts in twilight vision to match the blue light scattered from the sky after the sun has set. The storage time of the eye is closely matched to the speed of response or reaction time of the rest of the human system. The mechanical fineness of the rod and cone structure of the retina is internally self-consistent with the diffraction limits set by the diameter of the eye pupil. Perhaps one can decipher in the particular choice of rod and cone size, something of the primitive life habits of man. The hawk makes good use of an even finer structure in tracking down its prey from great heights. In fact, a most fascinating account of the great variety of adaptations of the visual system to living habits is given by Detwiler [7].

A sense of the efficacy of evolutionary processes is conveyed by the fact that no single man-made device yet matches the eye for compactness, versatility, and high performance throughout seven decades of light intensity.

REFERENCES

1. E. BAUMGARDT, Anée Phychol., **8**, 431 (1953)
2. H. R. BLACKWELL, J. opt. Soc. Am., **36**, 624 (1946)
3. E. M. BRUMBERG, S. I. VAVILOV and Z. M. SVERDLOW, J. Phys. (U. S. S. R.), **7**, 1 (1943)
4. A. D. COPE and A. ROSE, J. appl. Phys.. **25**, 240 (1954)
5. J. W. COLTMAN, J. opt. Soc. Am., **44**, 234 (1954)
6. H. J. DARTNALL and C. F. GOODEVE, Nature, **39**, 409 (1937)
7. S. R. DETWILER, Am. Scient., **44**, 45 (1956)
8. H. L. de VRIES, Physica, **10**, 553 (1943)
9. H. L. de VRIES, Rev. opt., **28**, 101 (1949)
10. H. K. HARTLINE, J. opt. Soc. Am., **30**, 239 (1940)
11. S. HECHT, Proc. natn. Acad. Sci. U. S. A., **23**, 227 (1937)
12. S. HECHT, J. opt. Soc. Am., **32**, 42 (1942)
13. B. KAZAN and F. H. NICOLL, Proc. Inst. Radio Engrs., **43**, 1888 (1955)
14. M. KELLER and M. PLOKE, A. angew. Phys., **7**, 562 (1955)
15. M. LUCKIESCH and A. H. TAYLOR, Illum. Eng., **38**, 4 (1943)
16. J. D. McGEE, Proc. Inst. Elec. Engrs., (Lond.), **97**, 377 (1950)
17. W. V. MAYNEORD, Brit. J. Radiol., **27**, 309 (1954)
18. G. A. MORTON, J. RUEDY and G. KRIEGER, RCA Rev., **9**, 419 (1948)
19. W. J. OOSTERKAMP and J. TOL, 7th Int. Conf. Radiology, Copenhagen, (1954)
20. R. K. ORTHUBER and L. R. ULLERY, J. opt. Soc. Am., **44**, 297 (1954)
21. P. REEVES, J. opt. Soc. Am., **4**, 35 (1920)
22. A. ROSE, J. Soc. Motion Picture Engrs., **47**, 273 (1946)
23. A. ROSE, J. opt. Soc. Am., **38**, 196 (1948)
24. A. ROSE, Adv. Elect., **1**, 131 (1948)
25. W. A. H. RUSHTON and R. D. COHEN, Nature, **173**, 301 (1954)
26. O. H. SCHADE, RCA Rev., **9**, 5 (1948)
27. O. H. SCHADE, J. opt. Soc. Am., **46**, 721 (1956)
28. R. E. STURM and R. H. MORGAN, Am. J. Roentgenol. Radium Therapy, **62**, 617 (1949)
29. G. SZIKLAI, R. C. BALLARD and A. C. SCHROEDER, Proc. Inst. Radio Engrs., **35**, 862 (1947)
30. S. A. TALBOT, Fed. Proc., **4**, 70 (1945)
31. R. THEILE, Arch. Electrisch. Übertrag., **7**, 15 (1953)
32. T. TOL, W. J. OOSTERKAMP and J. PROPER, Philips Res. Rep., **19**, 141 (1955)
33. H. A. van der VELDEN, Physica, **11**, 179 (1944)
34. G. WALD, Science, **119**, 887 (1954)
35. P. K. WEIMER, S. V. FORGUE and R. R. GOODRICH, Electronics, 70, May, 1950
36. R. CLARK JONES, J. Opt. Soc. Am. **49**, 645 (1959) and H. B. BARLOW, J. Physiol. 160, 155 (1962) (for measurements of quantum efficiency of the eye)
37. R. P. CHAMBERS and J. S. COURTNEY-PRATT, Phot. Science and Engineering, **13**, 286 (1969) (for measurements of threshold signal-to-noise ratio)

III. Television Pickup Tubes and the Problem of Vision

I. INTRODUCTION

The visual process in its most refined state is a simple counting process. For a given exposure time, scene brightness, and optical system, a finite number of light quanta will be radiated from the scene and absorbed by the particular device that looks at the scene. The device counts these quanta and from their number and distribution determines how many discrete bits of information it can furnish. The arithmetic used to convert numbers of quanta into numbers of bits of information will be outlined in a later section. For the present an attempt will be made to contrast the almost endless variety and complexity of picture pickup devices with the simplicity and universality of the performance scale according to which any of these devices may be judged. Throughout this paper emphasis is placed on performance limited by the finite number of available light quanta. This is patently a fundamental limitation. Performance limited by electron optical or structural defects will receive only passing mention. The technical problems involved in removing these defects may indeed be the most important and the most difficult for the success of a particular device. They are, nevertheless, removable defects and not fundamental limitations.

II. MAJOR TYPES OF PICKUP DEVICES

To name television pickup tubes, photographic film, and the human eye is to name examples of the three major types of visual processes: electrical, chemical, and biological. The detailed mechanics of the operation of a television pickup tube has little in common with that of photographic film and even less, perhaps, in common with that of the human eye. This, in spite of literary references to television pickup tubes as "electric eyes" and in spite of pictorial explanations of the human eye confined to the parallelism of parts — lens, black box, and film — of an ordinary camera. What is common to all of these devices, especially if they are well designed, is some means for counting incoming light quanta. The particular means for, or mechanics of, counting may be as varied as the imagination of man and nature combined. Television tubes generally convert light quanta into photoelectrons and count the number of electrons by measuring a current or a voltage. Photographic film, by an intermediate chemical process, converts each quantum (or at most a small

Reprinted from: Advances in Electronics, Vol. 1, 1949, pp. 131–166. Academic Press Inc., N.Y.

number of quanta) into an opaque granule of silver and observes the density of the silver deposit as a measure of the number of granules. The human eye, also, by an obscure intermediate process, converts a small number of incoming quanta into a single nerve pulse. The brain estimates the number of original quanta by the frequency of arrival of these nerve pulses and (a speculative thought) by the regularity of their arrival.

A television pickup tube differs again from the photographic process and from human vision by virtue of having to arrange its information (or picture) in a form convenient for transmission to a remote point. The "arranging" is done by a scanning process in which all of the bits of information in any one picture are strung end-to-end and passed single file through the television transmitter. At the television receiver, a converse arrangement, also a scanning process, accepts the bits of information and distributes them one at a time into their proper places to re-form the picture. Each picture is composed of a few hundred thousand bits of information or picture elements. And 30 pictures are transmitted/second. Because the persistence of vision of the eye is a few tenths of a second, neither the point-wise assembly of any one picture nor the discrete succession of pictures is resolved. Rather, the eye sees a smooth flow of events as in the original scene.

The scanning process dictates its own special problems and in some ways complicates the design of television pickup tubes. The scanning process does not, however, affect the performance attainable by these tubes as compared with the performance attainable by devices that do not use scanning.

III. NUMBER AND VARIETY OF TELEVISION PICKUP TUBES

Human vision, the photographic process and television not only compose the three broad approaches to the problem of seeing but also point up the variety of solutions to this problem. An even greater variety, numerically at least, is found within the television approach itself. Here, the patent literature offers literally hundreds of examples of ingenuity of design applied to television pickup tubes. Many more examples can be added to these by purely engineering combinations of the already patented devices.

With no intent of making the formula exhaustive, a large number of pickup tubes may be generated in the following fashion. Trace the steps in the formation of a picture. To each step assign a number equal to the number of ways that step may be performed. Take the product of all of the step numbers. The product, so formed, is an upper limit to the number of different pickup tubes that may be designed. Because not all of the ways of performing each step are known, this operation is more likely to yield a lower limit than an upper limit. The operation will be carried out for its suggestive value at this point. The coverage will be sufficient only for the purpose of illustration.

The steps in the formation of a picture are these. Incoming light quanta are converted into something that can be stored and counted. An element, usually called a target or mosaic, is provided capable of storing

the photoproducts for a thirtieth of a second. The target is scanned by some means that can "count" the photoproducts at each point on the target. The scanning process may return the target to its previous unexposed state or not. If it does not, a separate erasing mechanism is needed so that cyclic operation is possible. The "count" made by the scanning means is passed on to an amplifier and through other circuits to the television transmitter. The amplifier has a background "count" or noise of its own which tends to obscure the "count" that is fed into it. For this reason it is highly desirable to have some noiseless amplifier within the pickup tube that can magnify its "count" to a conveniently high level before it is fed into the external amplifier. Five steps have just been enumerated: Conversion, storage, scanning or counting, erasure, and noiseless amplification. Illustrative examples of ways of performing each step will be cited.

Light quanta may be converted into photoelectrons, excited electrons, ions, excited atoms, or dissociated molecules. The photoelectrons in turn may be accelerated and focused on to a target to excite other electrons or to dissociate or ionize molecules. The target used for storage may be one sided or two sided, insulating or semiconducting, continuous or apertured. The target may be scanned by a beam of electrons or a beam of light. The electron scanning beam may be of the low, medium or high velocity types, each velocity range being qualitatively distinct. The scanning beam may simply strip off the charge pattern deposited on the target by the picture, or it may charge the target to some arbitrary uniform potential. The scanning beam may not even strike the target, in which case only the paths of the electrons in the scanning beam are controlled by the charge pattern. If the scanning process has not restored the target to the unexposed state, a separate means such as a steady electrical leakage, a uniform electron spray, or a second scanning beam may serve the purpose. Finally, the scanning beam, after it has been influenced by the target, can be amplified, substantially without distortion, by an electron multiplier. (Alternatively, electron multiplication could have taken place in the previous process of laying down the charge pattern on the target.) In those cases where the beam does not strike the target, but is only controlled by it, large scanning beam currents may be used that do not require further amplification before being fed into the external amplifier. Without actually assigning the previously mentioned step numbers, one can see that even this rapid and incomplete enumeration will lead to hundreds of different pickup tubes.

If the above recital of ways of making pickup tubes has been more confusing than informative, it was not without plan. The present paper in no way aims at being a manual for designing pickup tubes nor does it attempt a critical survey of the various proposals that have been made or could be made for these tubes. Such a paper would indeed have an extraordinarily small "public. " The recital was introduced to underline the highly un-unique character of any one pickup tube — and there are scores of them. It was introduced as a self-evident argument for singling out only a few pickup tubes for close attention and these for the one purpose of tracing the successive realized approximations to ideal pickup tube performance. The chronological order turns out to have also a certain logical sequence.

IV. COMPARISON OF ACTUAL AND POSSIBLE PICKUP TUBES

In the opening sentence of this paper the visual process was described as a simple counting process. Conceptually it is easy to demonstrate both that the maximum, or ideal, performance is achieved insofar as each incoming quantum can be counted, and that such a count bears a simple relation to the intelligence that can be transmitted. Conceptually, also, one can design a simple pickup tube to satisfy this definition of ideal performance. It is perhaps something of a disappointment to find that the pickup tubes that have come into use are generally not simple. Either their construction is elaborate or their detailed operation difficult to analyze. What is more, they have still not exhausted the possibilities of ideal performance. This situation is not unusual in the early stages of an art, and television has yet to be matured in the atmosphere of a successfully commercialized system — that the first attempts are not only imperfect but elaborately imperfect. Another circumstance that restricts the freedom with which promising designs for pickup tubes can be plucked out of a file and converted into useful instruments is the number of side conditions such tubes must satisfy. These conditions apply largely to the target. Uniformity of response, for example, is one of the most difficult to satisfy. It is the problem of getting a hundred thousand little elements to act alike. Another side condition is that remnants of pictures shall not be noticeable in succeeding pictures, that is, that the process of erasure be substantially complete. Resolution in excess of the circuits is an obvious but not an easily met condition. Freedom from spurious signals is a condition that has ruled out many otherwise promising experimental tubes. Finally, there are reasonable limits to the kinds of targets that can be fabricated, at least without extensive and costly research. And the paper designs for targets are well ahead of the art of making them.

These considerations are meant not to justify the complexity or imperfection of present tubes but rather to give a measure of the difficulties of realizing the simplicity and perfection inherently possible in pickup tube design. The simple counting process that forms the basis for ideal performance will be outlined in the next two sections both analytically and experimentally.

V. IDEAL PERFORMANCE

The object of this section is twofold. First, a completely general relation will be derived for the amount of information a picture pickup device can furnish as a function of the brightness of scene at which it is directed. Ideal performance will be insured in this derivation by the assumption that the pickup device can count each absorbed quantum of light. Second, in so doing, a general and absolute scale of performance will be set up according to which the performance of actual pickup devices may be judged. Use will be made of this scale of performance in a later section.

Consider a square element of area of side length "h" on the photo-sensitive surface of the picture pickup device. Let this element of area absorb on the average N quanta in the exposure time alloted. Because the absorption of quanta is a random process, the average absorption number, N, will have associated with it deviations from the average whose root mean square value is $N^{\frac{1}{2}}$. These deviations set a limit to the accuracy with which the average number N may be determined. By the same token they set a limit to the smallest change in N that may be detected. Let this smallest change in N be denoted by ΔN. Thus, ΔN is of the order of $N^{\frac{1}{2}}$. The particular constant of proportionality will depend from probability considerations on the certainty asked for in detecting ΔN. With the above definitions, several relations can be written out of hand.

$$\text{Scene brightness} = B \sim \frac{N}{h^2} \tag{1}$$

$$\text{Threshold contrast} = C = \frac{\Delta B}{B} = \frac{\Delta N}{N} \sim N^{-\frac{1}{2}} \tag{2}$$

$$B \sim \frac{1}{C^2 h^2} \tag{3}$$

or

$$BC^2 h^2 = \text{constant} \tag{3a}$$

or

$$BC^2 \alpha^2 = \text{constant} \tag{3b}$$

where α is the angle subtended by "h" at the lens.

Eq. (3b) is already the characteristic equation of the ideal pickup device. The constant term on the right will be evaluated shortly. It contains in straightforward fashion, the lens parameters, exposure time, and quantum efficiency of the device. The meaning of Eq. (3b) is this. Let any two of the variables (B, C, α) be specified. Eq. (3b) then sets the threshold value for the third. For example, if the scene brightness and contrast are specified, Eq. (3b) gives the smallest angular size that can be resolved. Conversely, if one knows the angular size and contrast of a test object, Eq. (3b) determines the minimum scene brightness required for detecting it. It is to be noted that the resolution (α) or halftone discrimination (C) improve only as the square root of the scene brightness. It is to be noted also that the validity of Eq. (3b) rests only on the countability of the absorbed quanta and the random character of the absorption process. The first was the necessary condition for ideal performance; the second is a well established experimental fact. Since no special mechanism was assumed, Eq. (3b) can apply equally to the eye, to photographic film, and to the great variety of television pickup tubes already discussed insofar as each of these devices can make an accurate count of the absorbed quanta.

Eq. (3b) is the basis for a performance scale according to which the performance of any particular device, whether it be eye, film, or pickup tube, may be judged. It is a relatively straightforward operation to measure the smallest angle that a given device can resolve at a specified scene brightness and contrast. The product of these three quantities, as in Eq. (3b) yields a number whose value locates the position of the given device on the performance scale. This scale is an absolute one in the sense that performance is limited only by the statistical fluctuations in the rate of absorption of light quanta, a fundamental and unavoidable limitation. Whatever happens subsequent to the primary process can only deteriorate, or at best maintain, the performance already computed in terms of the primary process. Subsequent elements can not improve performance.

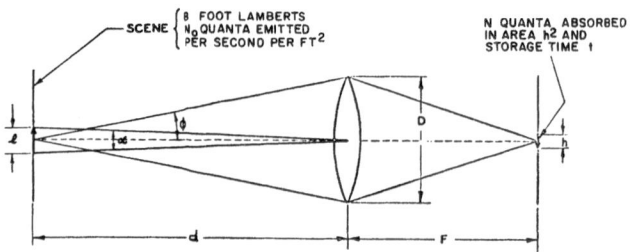

FIGURE 1. Geometric relations used to compute number of absorbed quanta as a function of number of emitted quanta.

The performance scale, based on Eq. (3b), will be more significant if the constant term on the right is separated into its component factors. To do this, reference is made to Figure 1. Consider an element of side length "1" in the scene. If the scene brightness is such that N_0 quanta are radiated/square foot/second, the total number of quanta radiated by this element per second will be $N_0 1^2$. Of this number, only a fraction will be intercepted by the lens and, for a Lambert distribution, the number passing through the lens per second is:

$$N_0 1^2 \sin^2 \phi$$

This number of quanta will also be incident on the element of side length "h" which is the image of the scene element of side length "1" The number of quanta actually absorbed at "h" in the exposure time of the device is

$$N = \theta t N_0 1^2 \sin^2 \phi$$

where θ is the quantum yield or absorption coefficient* and t is the exposure time. From the relations (see Figure 1):

$$1 = \frac{d}{F} h$$

$$\sin \phi \doteq \frac{D}{2d}$$

and

$$\alpha \doteq \frac{h}{F}$$

_____

* The absorption coefficient is restricted to those quanta that give rise to countable events.

Eq. (4) may be rewritten in the form

$$N = 1.4N_0 D^2 t \theta \alpha^2 \times 10^{-10} \tag{5}$$

where α is expressed in minutes of arc, and D is expressed in inches. Using the equivalence: 1 lumen of white light = 1.3×10^{16} quanta/sec,

$$\frac{N_0}{1.3 \times 10^{16}} = B \text{ ft.-L}$$

and, from Eq. (5)

$$N = 2BD^2 t \theta \alpha^2 \times 10^6$$

or

$$B = 5 \frac{N}{D^2 t \theta \alpha^2} \times 10^{-7} \text{ ft.-L} \tag{6}$$

From Eq. (2), the threshold contrast is related to $N^{-\frac{1}{2}}$ by a constant whose value is yet to be determined. This constant is the threshold signal-to-noise ratio and, as mentioned, is a function of what one calls threshold. That is, if one asks that each observation have a 90% chance of being correct, the threshold signal-to-noise ratio will be higher than if one were satisfied with a 50% chance. From visual observations to be cited later, a reasonable value for the threshold signal-to-noise ratio appears to be greater than unity and probably in the neighborhood of 5. For the present this constant will be introduced by the letter "k." Thus, inserting k^2/C^2 for N in Eq. (6) we get the complete form of the characteristic equation,

$$BC^2\alpha^2 = 5 \frac{k^2}{D^2 t \theta} \times 10^{-7}$$

or, expressing C as a per cent contrast $\left(C = \frac{\Delta B}{B} \times 100 \right)$ we get:

$$BC^2\alpha^2 = 5 \frac{k^2}{D^2 \theta} \times 10^{-3}$$

An inspection of the right hand side of Eq. (7) shows that the only parameter that differentiates the performance of one ideal pickup device from that of another is the quantum yield (θ) of the primary photo-process. This statement will probably be clearer if Eq. (7) is rewritten in the following form:

$$\frac{5k^2}{BC^2\alpha^2 D^2} \times 10^{-3} = \theta \tag{8}$$

The five parameters needed to specify a given set-up are in the denominator of the left hand side of Eq. (8). Their names and units are: scene brightness (B ft.-L), contrast of test object $\left(C = \frac{\Delta B}{B} \times 100\% \right)$, angular size of test object (α minutes of arc), lens diameter (D in.) and exposure time (t sec). When any four of these are arbitrarily specified, the threshold value for the fifth is given by Eq. (8). Alternatively, if all five parameters are known, a value for the quantum yield (θ) is thereby determined. If the operation of the device is ideal, that is, if all of the absorbed quanta are counted, the value for θ determined in this way is actually the quantum yield of the primary photoprocess. If, on the other hand, the operation departs from ideal operation — and the ways in which such departures may occur are both manifold and frequent — the value for the quantum yield so determined is l e s s than the quantum yield of the primary photoprocess. In this event, it may still be looked upon as an index to performance. To recapitulate: a given value for θ computed from Eq. (8) may correspond to an ideal device having the same value of θ for its primary photoprocess or to a nonideal device having a larger value of θ for its primary photoprocess, the nonideal characteristic being responsible for the lower c o m p u t e d value of θ.

Eq. (8) defines the absolute performance scale to be used later in evaluating the performance of various pickup tubes, photographic film, and the eye. The scale extends from $\theta = 0$ to $\theta = 1$. The value $\theta = 0$ means no transmitted picture. The value $\theta = 1$ means not only ideal operation but absorption of all incident quanta as well. It should be clear at this point that a single value of θ may be used to designate the performance of a device only over a limited domain of the five parameters B, C, α, D and t. In fact, in the usual case, θ is a continuous function of these parameters. No device, for example, can resolve arbitrarily small angles. Nor can any device discriminate arbitrarily small contrasts. Nonstatistical considerations set limits to the smallest values attainable by either of these parameters. Also, the range of times over which a device can store up information has an upper limit set by the mechanics of the device and not by statistical limitations.

It would indeed be a thorough-going paper that explored the performance of various pickup devices for variations of all five of the parameters of Eq. (8). Such completeness will not be attempted here. Rather, only one of the parameters, the scene brightness, which is of greatest interest, will be selected for examination. Reasonable fixed values or ranges of values of the other parameters will be taken. Some remarks will be made in passing about the dependence of θ on these other parameters but with no object of making the review exhaustive. The parameter, scene brightness, is of special interest because few devices can maintain a high level of performance over an appreciable range of scene brightnesses. The eye is, perhaps, the outstanding example. Other devices may match or even exceed the eye over narrow ranges of scene brightness but, as yet, they lack the flexibility of the eye to maintain performance both at very high and at very low lights.

Before estimating the performance of various selected pickup devices, it will be well to try to tie the characteristic Eq. (8) closer to reality, both by inserting specific values and by outlining an experimental arrangement which satisfies this equation in simple fashion. This arrangement will also furnish an experimental value for threshold signal-to-noise ratio, k.

In Figure 2, the threshold contrast has been plotted against the reciprocal angular size for a large range of scene brightnesses. The values for quantum efficiency, storage time, and lens diameter shown in this figure were selected to approximate the human eye over part of its range of operation. The value 5 was assumed for the threshold signal-to-noise ratio (k) in computing these curves. The dash-dot lines indicate the limits to the smallest angular size and contrast that can be detected, which limits are set variously by the particular mechanisms of particular devices and are not set by statistical fluctuation considerations. For this reason, the curves are dotted outside the dash-dot boundaries.

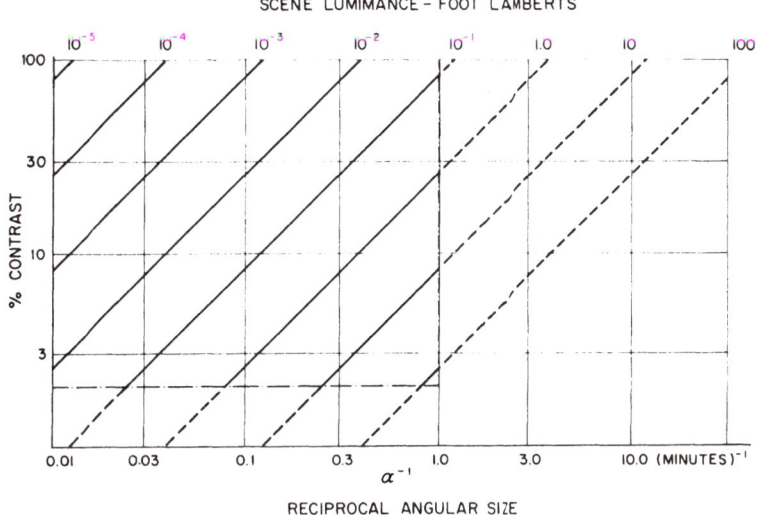

FIGURE 2. 'Performance curves for an ideal pickup device.

By way of example, Figure 2 shows that, at a scene brightness of 1 foot-lambert, an ideal device having a quantum yield of 0.01, a storage time of 0.2 sec and a lens diameter of 0.3 in. can just detect a contrast of 30% for objects subtending 1 min of arc at its lens. If smaller contrasts, say around 3%, are to be detected, the object must be larger and subtend an angle of about 10 min. At very low scene brightnesses, Figure 2 shows that only a poor quality picture may be transmitted (poor resolution and poor discrimination for half tones) corresponding to the obscured pictures that the eye sees at night. The poor quality of these pictures is a direct consequence of the lack of sufficient light quanta and cannot be avoided by any "improvements" in design other than improved quantum yield in the primary photoprocess.

Lest the present discussion be dominated by a purely speculative tone, an actual physical system will be described in the next section which brings out the essential features of ideal performance.

VI. AN EXPERIMENTAL REALIZATION OF IDEAL PERFORMANCE

An early and simple means for generating a television picture is embodied in the so called light-spot scanner. Figure 3 shows the essential parts of this system. The scene to be transmitted is "illuminated" by a small sharply focused spot of light which scans the scene in the customary manner in a series of parallel lines. At each point in the scene some of the scanning spot light is reflected and is picked up by a photocell. The light and shade of the scene are conveyed by the variations in scattered light as the spot scans over its surface, and the variations in scattered light are in turn conveyed by variations in photocurrent in the photocell. A viewing tube or kinescope is connected to the photocell in such a way that, as the original light spot scans its scene, the kinescope beam scans its luminescent screen in synchronism and, by variations in beam current controlled by the output of the photocell, reproduces the original scene. The light-spot scanner is restricted to those scenes that may be conveniently illuminated by a scanning spot of light. Recent developments in phosphors and photomultipliers have revived interest in the light-spot scanner for those applications [1]. Its operation is simple and it provides a picture free from most of the spurious signals frequently encountered in storage type pickup tubes.

The particular virtue of the light-spot scanner for the present discussion is that it offers a ready means for demonstrating the properties of ideal performance. For example, the gain of the electron multiplier in the photocell is sufficiently high that each quantum that is absorbed at its photocathode (and liberates a photoelectron) can be made visible on the kinescope as a discrete speck of light. Thus, at extremely low scene brightnesses, one can actually and easily count the number of "quanta" in the reproduced picture.

The special test pattern which served as subject for the light spot scanner and which measured its performance is shown in Figure 4. This pattern is a materialization of the plot in Figure 2. The discs along any row have the same contrast but vary in diameter step-wise by a factor of 2 for each step. The discs in any column have the same diameter but vary in contrast also step-wise by a factor of 2 for each step. At moderate or low illumination not all of the discs in this pattern can be seen or transmitted by a pickup device. The demarcation between those discs that can be seen and those that cannot

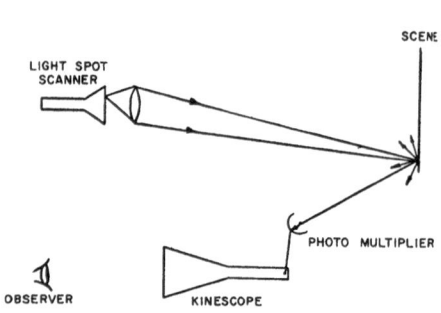

FIGURE 3. Light-spot scanner arrangement.

should be a 45° diagonal if the picture is limited only by noise whose distribution in frequency is uniform (e. g., shot noise in a temperature limited beam of electrons). As the illumination of the test pattern is varied the line of demarcation should move from one diagonal of discs to the next for a factor of 4 change in illumination.

The series of photographs in Plate III * is a series of timed exposures of the kinescope taken as the light-spot scanner scanned the test pattern at the usual television rate of 30 pictures/sec. For technical convenience the exposure time of the camera was varied rather than the scene brightness since, according to Eq. (8), it is only the product Bt that is significant.

The exposure times in the first few pictures are low enough to see the separate "quanta" (or their corresponding specks of light on the kinescope). As the exposure time is increased, more and more of the test pattern becomes visible. In particular, for a range of over 200 in exposure times, the demarcation shifted one diagonal of discs for each factor of 4 increase in time. Also, the demarcation on any one picture is approximately along a 45° diagonal. These two observations confirm the significant features of Eq. (8), namely that at any one scene brightness, threshold contrast is proportional to the reciprocal angular size and that both the reciprocal contrast and the reciprocal angular size vary as the square root of the scene brightness.

The series of photographs in Plate III provide also a means for estimating the exposure time of the eye and the value of k, the threshold signal-to-noise ratio.

Concerning the exposure time of the eye, it was found that, of the series of photographs of varying exposure times, the picture that most nearly matched the direct visual impression of the kinescope was that for 0.2 sec. This test was made actually more quantitative by placing a mask over the kinescope and counting the average number of specks visible in a small area. The particular photograph was then selected that had the same average number of specks in the same area. The exposure time of the eye determined in this manner is in satisfactory agreement with well known and more thorough determinations already published.

Concerning the threshold signal-to-noise ratio, "reasonable" assumptions, for lack of experimental evidence, have been made [2, 3] that its value is unity. A recent direct determination [4], however, sets the value of k in the range of 3-6, depending on the viewing conditions. The present paper has used the value 5. Whatever uncertainty there may be in estimating a value for the threshold signal-to-noise ratio from the photographs in Plate III, that value is definitely greater than unity and is in the range of 3-6. This is surprisingly high since mathematical analysis suggests a value close to unity**. One way of reconciling the discrepancy is to assume that the eye in looking at the pictures in Plate III does not make full use of the information presented. That is, the eye (and brain) may count only a fraction of the "quanta" present. If one used a machine to do the counting and arranged to ring a bell whenever the count warranted, one might approach a threshold ratio of unity. Since, however, one usually chooses to do his own counting by the normal visual process, the experimentally determined value of 5 for threshold signal-to-noise ratio is significant.

---

*Plates II and III are located on pp. 87−88.

** Note added in proof: These statements have meaning for viewing a single picture element by itself but not for viewing a whole picture consisting of many picture elements. They do not take into account the presence of statistically spurious signals discussed in the preceding paper. An analytic argument is presented there for choosing k to be approximately 5.

The operation for computing a threshold signal-to-noise ratio from Plate III is as follows. Select the smallest black (not grey) disc that is visible in any one of the photographs in which the specks are countable. Transpose the outline of the black disc to the surrounding uniform white area. Count the average number of specks in the outlined area. This average number is the signal; the square root of the average number is the root mean square noise. The threshold signal-to-noise ratio is then also the square root of the average number. A similar operation can be carried out to get the same threshold signal-to-noise ratio using the outline of a threshold grey dot. In this instance the signal is the average number multiplied by the contrast (not per cent contrast) and the noise is the square root of the average number. The results of these operations, as already mentioned, lead to a value of k in the neighborhood of 5.

The light-spot scanner was used also to make a direct comparison estimate of the performance of the eye. Even though the light-spot scanner is a nonstorage pickup device and has the characteristic insensitivity of such a device, it turns out that the average illumination on the scene as provided by the scanner is as low as it would be for an ideal storage type pickup tube operating with uniform continuous illumination. The reason is that the scanner illuminates the scene only when and where that illumination can be used. There is no waste illumination. Hence, if one measures the average scene brightness for a scene illuminated by a light-spot scanner and observes the signal-to-noise ratio of the transmitted picture these quantities correspond exactly to the performance of an ideal pickup device having the same quantum efficiency as the photosurface of the photomultiplier. To complete the parallel, the diameter of the photocathode of the multiplier replaces the diameter of the lens in Eq. (8). The exposure time of the system is the exposure time of the observer (human or instrumental) that looks at the final picture on the kinescope. The direct comparison between photomultiplier and eye was further facilitated by the fact that the useful area of the multiplier photocathode is about equal to the diameter of the dark-adapted eye and that the spectral distribution of the scanner light overlapped about equally the response curves for eye and multiplier. The test was carried out by having the photomultiplier and human eye both look at the original scene at the same viewing distance. The human observer then looked at the kinescope to compare what the photomultiplier saw with what the human eye could see directly on the original scene. Using the test pattern of Plate II it was interesting to find the eye and photomultiplier performance very closely the same in the test range of 10^{-2} to 10^{-4} ft.-L. Since the picture transmitted by the photomultiplier was obviously limited by fluctuation noise in its primary photoprocess and since the quantum efficiency of the photomultiplier surface was about 2%, the quantum efficiency of the eye must also have been 2% if it were limited by fluctuation noise in its primary photoprocess or greater than 2% if some less fundamental limitation were present.

VII. PERFORMANCE OF SELECTED PICKUP DEVICES

The object in the following sections will be to trace the performance of various representative pickup devices as a function of scene brightness, using the absolute scale of performance already discussed. Because the range of scene brightnesses to be considered is over 10 orders of magnitude and the range of performance values (θ) over seven orders of magnitude, the precision attainable in a single composite plot will not be large. What is aimed at in this survey and the corresponding plot is a picture of the areas of performance not yet covered by any device as well as the areas already covered by the eye but not yet by any man-made device. In brief. it is a picture of performance possibilities yet to be realized*.

1. Human Eye

To locate the performance of the eye, the six parameters on the left hand side of Eq. (8) need to be known. There are data in the literature [5, 6, 7] relating scene brightness (B), threshold contrast (C) and angular size (α) for the eye. These data cover the following ranges: B, 10^{-6}-10^2 ft.-L, C, 1-100% contrast, and α, 1-100 min of arc. In addition, less complete observations have been made above 10^2 ft.-L and below 10^{-6} ft.-L. For example, above 10^4 ft.-L pain sets in and below 10^{-7} ft.-L vision ceases completely. There are also data for the pupil diameter (D) as a function of scene brightness [8]. If the storage time (t) of the eye is taken to be 0.2 sec, and independent of scene brightness, and if, for the threshold signal-to-noise (k), the previously discussed experimental value of 5 is taken, the roster of information necessary to compute performance (θ) is complete. One further qualification needs to be made. It is found from Blackwell's data that at any fixed value of scene brightness, the threshold contrast is only approximately proportional to the reciprocal angular size as demanded by Eq. (8). One can, however, for the purpose of this plot, select the best performance at each value of scene brightness (see also [9]). This procedure was used to compute the curve of θ vs. B for the eye in Figure 4.

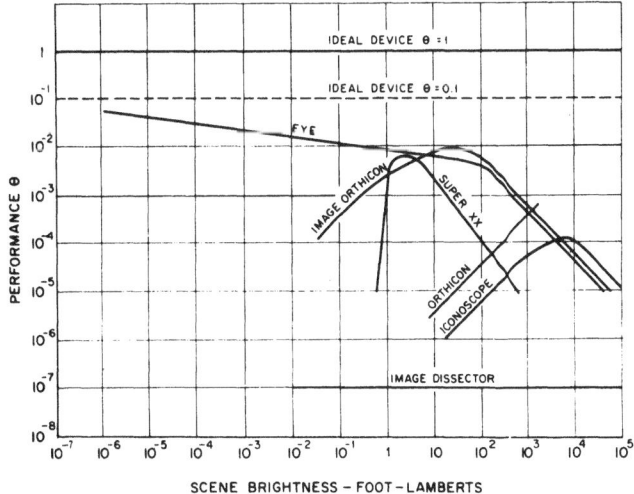

FIGURE 4. Performance curves for selected pickup devices.

---

* Since this paper was written, the Image Intensifier Orthicon (see review by P. K. Weiner) has matched most of the low light end of the eye curve.

The most striking feature of this curve is the exceedingly large range of scene brightnesses, namely 10^{-6} to 10^2 ft.-L, over which the eye maintains a high level of performance. It is striking because no single man-made device has succeeded in covering this range. Even more, the ensemble of man-made devices falls far short of covering it. The experience of those who have tried to build the same flexibility and excellence of performance into electronic devices can only suggest the artfulness of the mechanisms that in some way have been engineered into or have selectively survived in the eye. Two of these mechanisms in particular deserve mention.

There is good evidence that the eye requires more than a single absorbed quantum to generate a visual sensation. If, now, the rods and cones had no interconnections but acted completely independently in their reception and transmission of information to the brain, the eye would rapidly fail to "see" when the concentration of absorbed quanta fell below the concentration of rods and cones. This, in fact, is just the reason that photographic film cuts off sharply towards low scene brightnesses. The eye, then, must have a mechanism that can pool the information from the separate rods and cones when such pooling is advantageous for seeing and can separate the operation of these rods and cones when, as at high lights, there is an abundance of light quanta. In brief, the mechanism must automatically control the interconnections of the retinal elements to get the maximum information out of the incoming stream of quanta.

The second interesting mechanism has to do with the existence of a variable-gain element located between the retina and the nerve fibers that carry pulses to the brain. A gain element of some sort is needed in order to raise the energy level of the few quanta required for a threshold sensation up to the much higher energy level of the subsequent nerve pulse. If the gain of the element were constant, however, the nerve fibers would be called upon to pass visual sensations varying (according to Figure 4) by about eight orders of magnitude. While such a design is not inconceivable, it is not especially elegant. By way of example, if one wants a device capable of recording the variation of a quantity in ten thousand discrete steps, it is simpler to compose the device of two elements each capable of one hundred different values, such that their product yields the ten thousand discrete steps than to try to design the ten thousand steps into a single element. Whatever the validity of this argument, there is good reason to believe from observations on dark adaptation, that the gain of the above mentioned element varies automatically with scene brightness [9]. [9]. The delay in "resetting" during the transition from high lights to low The delay in "resetting" during the transition from high lights to low lights would correspond to the time taken for dark adaptation. Also, fluctuations (noise) in the primary photoprocess (absorption of light quanta at the retina) should be visible if the computed value of (Eq. 8) is equal to the value of observed or computed directly from the retina. Hecht [10], for example, by a statistical analysis of visual observations near threshold arrives at a value of about 0.1. Hecht's value for blue light should be divided by a factor of 3 for comparison with the white light observations of Figure 4. The agreement is close enough to suggest that the variable gain is automatically set to make fluctuations just visible. From necessarily subjective observations the writer confirms this conclusion at least at low lights around 10^{-4} foot-lamberts.

This discussion leaves out the mechanisms, probably still more ingenious, that are responsible for color vision.

2. Photographic Film

The millions of years spent in evolving the human eye have yielded a device with a high level of performance over an enormous range of light intensities. The one hundred or more years spent in the development of photographic film have resulted in the same level of performance but confined to a very narrow range of light intensities. With adequate light, pictures produced photographically are remarkable for their uniformity, resolution, rendition of tonal values, and freedom from distortions and spurious effects. As the scene brightness is reduced, however, a relatively sharp threshold is reached below which no photograph, not even a poor one, is obtained. This abrupt departure from ideal performance results mainly from the fact that more than one absorbed quantum is needed to make a photographic grain developable. When the density of absorbed quanta falls below the density of grains, the probability that more than one quantum will be absorbed by a grain rapidly approaches zero. If film could pool its grains, as the eye does its retinal elements, it also could continue to record pictures and maintain its performance level at low lights.

A given film departs from ideal performance, but more slowly, also towards high lights. This departure becomes clear if one first exposes a film so that about half its grains are rendered developable, that is, to a latent density of about 0.3. Now, if the film is exposed further, at least half of the new exposure is wasted on grains that have already been made developable. As the exposure continues, the waste increases and so does the departure from ideal performance. The finite number of grains, and correspondingly finite signal-to-noise ratio, further restricts the approach to ideal performance at high lights.

The relatively sharp threshold towards low lights and the progressive inefficiency of exposure at high lights leads to an optimum of performance for a given film in the neighborhood of a density of 0.3. This optimum can be shifted along the scene brightness scale (Figure 4) by choice of films with different grain sizes. A given film, however, for fixed values of lens speed and exposure time, has a rather narrow working range.

To locate the curve for Super XX film shown in Figure 4, data giving the measured signal-to-noise ratios as a function of area on the film and film density were used [11]. Also, a lens diameter of 0.3 in. and an exposure time of 0.2 sec were selected in order that the depth of focus and speed of operation be comparable with the same properties of the eye. A different choice of lens diameter or exposure time would not change either the shape of the curve or its vertical position on the performance scale. It would merely shift the curve rigidly along the brightness axis. Because the location of the point of departure from ideal performance at the low light end depends upon the relative density of light quanta and grains at the film, it is necessary also to specify the focal length of the lens. This insures a one to one correspondence between scene brightness and illumination at the film. The focal length is chosen to give a commonly used angle of view of 25°. Thus, for 35-mm film, the vertical height of whose picture is about 16 mm, the focal length would be 1.5 in. Another choice of focal length would, like variations in lens diameter or exposure time, only shift the curve rigidly along the brightness axis.

3. Television Pickup Tubes *

The following tubes have been selected from amongst a great variety of possible tubes, first because they have all been realized and put into service, and second because historically and logically they represent successive steps towards attaining ideal performance. A more detailed discussion of the mechanics of their operation can be found in the cited references, or in Zworykin and Morton [12], or in a recent summary [13].

To compute the pickup tube curves, a lens diameter of 0.3 in. and an exposure time of 0.2 sec were uniformly chosen for ready comparison with the eye. Also, the focal length of the lens was selected to give an angle of view of 25°.

It is interesting to note that the choice of 0.2 sec for the exposure time is not inconsistent with the usual television frame time of 0.03 sec. This longer exposure time is set by the final observer at the kinescope. The human eye, for example, integrates about six successive television frames in its storage time of 0.2 sec. The same holds for the viewing of motion picture film. The result is an improved signal-to-noise ratio over what one would have gotten if he actually observed separate frames.

It is well to point out also that for pickup tubes the focal length of the lens may in general be varied with no effect on the shape or position of their performance curves, providing the linear size of target is varied in the same proportion. This differs from film for which the intensity of illumination rather than the total illumination on the target is critical.

a. Image Dissector. The image dissector (Figure 5) [14, 15] rapidly attained the theoretical limit of performance for nonstorage type pickup tubes. Since, however, the lack of storage already puts this limit about five orders of magnitude (number of separate picture elements) below that for storage type tubes, the image dissector is confined to applications where sensitivity is not important. The transmission of motion picture film or slides is one such application.

The operation of the image dissector, except for its lack of storage, is an excellently simple example of the counting process that is the ultimate basis of all picture pickup devices. The picture to be transmitted is focused on a conducting photocathode. The photoelectrons are then focused at 1:1 magnification at the other end of the tube. Here the tube is closed except for a small aperture, the size of the desired picture element. The entire picture is deflected across the aperture so that the aperture can explore each portion of it every thirtieth of a second. The electrons that pass through the aperture are multiplied by an electron multiplier whose gain can be high enough to count individual electrons. At any one moment, the electrons from only one picture element pass through the apperture. If one counts the total number of electrons that pass

---

* See also P. K. Weiner, Advances in Electronics and Electron Physics, 13, 387 (1960) Academic Press, for the inclusion of more recent television pick up tubes.

through the aperture in the time it takes to scan a picture element, that number is a measure of the signal and its square root is actually the signal-to-noise ratio referred to a picture element. It is also the signal-to-noise ratio that would be measured in the conventional manner by observing the current and noise power with appropriate meters.

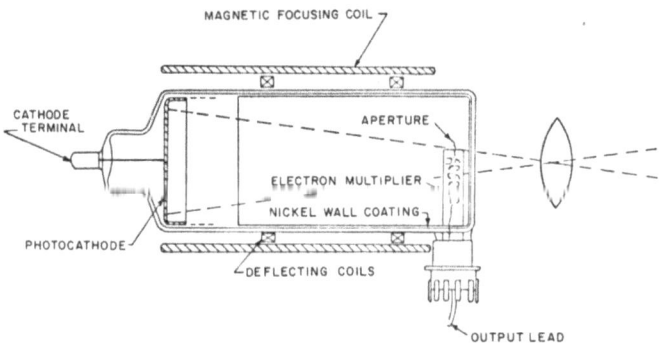

FIGURE 5. Image dissector.

In addition to the five orders of magnitude loss in performance due to lack of storage, the image dissector is down another two orders of magnitude (as are all the other pickup tubes*) because its photocathode has a quantum yield of about 0.01. The resulting curve in Figure 4 shows a constant level of performance over its full light range, but seven orders of magnitude down from the theoretical maximum.

b. Iconoscope. The iconoscope (Figure 6) [16] made the important step of introducing storage into pickup tubes. Although the iconoscope does not make full use of storage and is somewhat handicapped by a characteristic nonuniform shading pattern, it made possible the transmission of high quality "live" pictures. In fact, one of the very elements in its operation that is responsible for its inefficiency also leads to a reproduction of tonal values that has been difficult for other more sensitive tubes to match. This element is the incomplete collection of photoemission at high lights.

The optical image to be transmitted is focused on a photosensitive and insulating target. A charge pattern is stored which, when scanned by a constant current beam of electrons, modulates the escape of the secondary electrons. An amplifier is connected to the signal plate on the backside of the target and observes the variations in escape of secondary electron current. Equilibrium potentials are maintained by a rather involved process of redistribution of some of the secondary electrons over the target. While the saturated photosensitivity of the target may be about 10 microamperes/lumen, the operating photosensitivity is much less owing to lack of saturation. The operating photosensitivity is only 1 or 2 microamperes/lumen at low lights and a third to a tenth of that at high lights. Further, the noise associated with the picture is not the

---

* Modern photocathodes have quantum yields in the neighborhood of 0.1 corresponding to photosensitivities in the neighborhood of 100 microamperes per lumen.

noise in the escaping stream of secondary electrons but the much larger fixed noise characteristic of the amplifier. Thus, low-level signals (or low counts) are obscured proportionately more than high-level signals. For this reason, the performance curve (Figure 4) rises towards the high-light end.

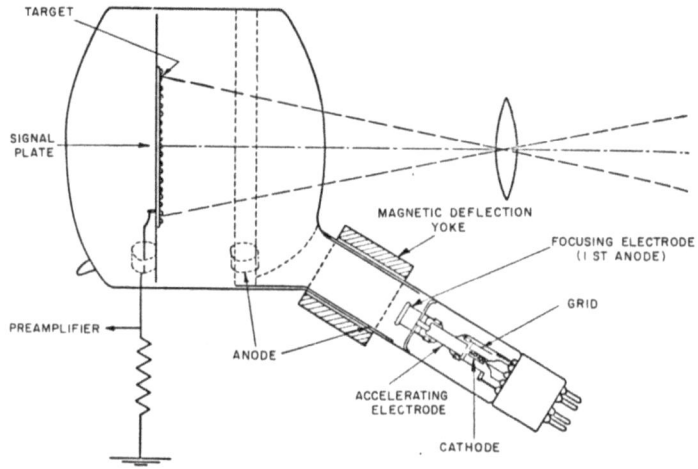

FIGURE 6. Iconoscope.

To compute the performance curve for the iconoscope, its known signal vs. light curve was used together with the noise current value (2×10^{-9} amperes) for a television amplifier with a pass band of 5 Mc. A more accurate appraisal would take into account the fact that the noise is not uniformly distributed over the pass band but is peaked at the high frequencies. The effective noise current may accordingly be as much as a factor of 3 lower [4].

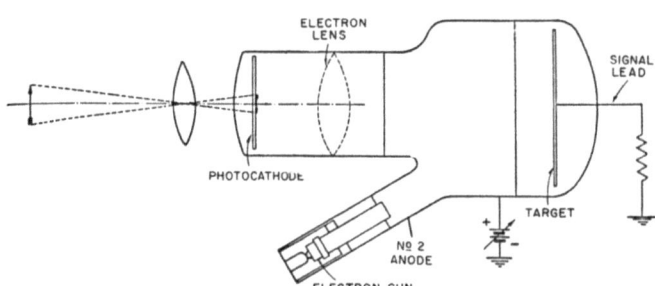

FIGURE 7. Image iconoscope.

The combination of a 0.3 in. lens diameter and a focal length of 9 in. (for 25° angle of view) leads to a lens speed of f/30 and also to the location of the iconoscope curve at exorbitantly high values of scene brightness. This is legitimate for the present purpose of showing what scene brightness the iconoscope would need if it were to match the depth of focus of the eye. In practice, however, the lens diameter is opened up at the expense of depth of focus in order to bring the operating curve down to reasonable values of scene brightness.

c. Image Iconoscope. The image iconoscope (Figure 7) [17] differs from the iconoscope by having an electron image rather than an optical image focused on the target. The electron image originates from a conducting photocathode on which the optical image is focused. Sensitivity gains are realized because the conducting photocathode can be made more sensitive than the insulating target surface and because the electron image is amplified by secondary emission at the target. The resultant sensitivity is five to 10 times that of the iconoscope. For the rest, its operation is substantially that of the iconoscope. Its performance curve lies in the immediate neighborhood of that for the orthicon, which is about to be described. Its curve is not included in Figure 4.

As mentioned earlier, the same limit of pickup tube sensitivity may be reached by multiplication of the electron image as by multiplication of the signal current delivered by the scanning beam. It has been technically more convenient, however, to get large gains by the latter process.

d. Orthicon. The orthicon (Figure 8) [18] avoided two prominent limitations of the iconoscope and at the same time introduced problems peculiar to its own design. The spurious shading pattern and the incomplete utilization of storage of the iconoscope both result from the complex redistribution of secondary electrons generated by the high velocity scanning beam. It was well recognized that a low velocity scanning beam would avoid these defects. Electrons from the scanning beam would then be deposited only where a positive picture charge pattern existed and in an amount equal to the positive charge. There would be little or no interchange of electrons between parts of the target. Also, a strong collecting field could be set up to saturate the photo emission from the target.

While the virtues of a scanning beam of low velocity electrons were known, it was also appreciated that such a beam was difficult to control. The beam could easily be defocused or deflected by stray charges in the tube and even by the picture charge on the target. Further, the beam had to be deflected in such a way that its angle of approach to the target was at least uniform, and preferably perpendicular, over all parts of the target. The orthicon, by its use of a uniform magnetic focusing field extending the full length of the tube, presented one useful solution to these problems. In the earlier tubes, the slow speed vertical deflection was accomplished by a pair of deflection coils while the high speed horizontal deflection made use of a pair of specially shaped plates. Improvements in deflection circuits made possible a later and simpler design in which both vertical and horizontal deflections were carried out with deflection coils.

As in the iconoscope, the optical image is focused on a photosensitive and insulating target surface. A positive charge pattern is thereby built up and stored. The scanning beam approaches the target at near zero velocity. Where no positive charge is present (dark parts of the picture) the beam reverses direction without striking the target. Where positive charge is present, some of the electrons in the beam land on the target and in an amount just equal to the positive charge. The semitransparent signal plate on the opposite side of the target records the fraction of the beam current that lands and passes this information on to a television amplifier.

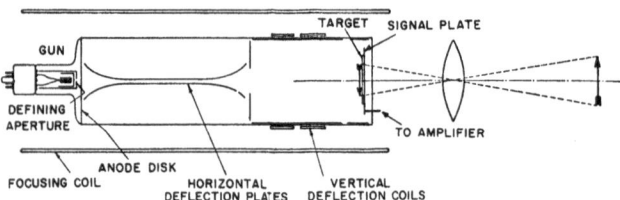

FIGURE 8. Orthicon.

The signal out of the tube is proportional to the incident light intensity. The curve extends from low signals whose threshold character is determined by the noise of the television amplifier up to a signal value more or less well defined by the need for maintaining picture quality. That is, at very high lights the scanning beam tends to be defocused and deflected by potential differences on the target.

The transmitted picture is free from spurious shading patterns. This fact, in combination with the linearity of the signal vs. light curve, has recommended the tube for picking up low-light, low-contrast scenes. On the other hand, difficulty has been encountered in trying to squeeze scenes with a wide range of contrasts into the limited signal range of the tube. Another operating problem results from the fact that the cathode potential of the target is a metastable potential. A sudden bright flash of light can charge the target up to the point where it is locked by the scanning beam at the relatively stable level of anode potential. A finite and objectionable amount of time is required to return the target to cathode potential.

To plot the orthicon curve in Figure 4 a target photosensitivity of 5 microamperes/lumen was assumed together with a focal length of 4.5 in. for the 0.3-in. diameter lens. The vertical height of the target is about two inches. The same television amplifier noise current was used as for the iconoscope to compute signal-to-noise ratios.

Mention should be made here of experimental orthicons that have been designed using electron multiplication of the return part of the scanning beam current. Higher sensitivities and signal-to-noise ratios have been obtained at some sacrifice of target stability at high lights.

e. Image Orthicon. The image orthicon (Figure 9) [19] has incorporated many of the useful operating characteristics of the image dissector, iconoscope, image iconoscope, and orthicon into one tube. It has the freedom from amplifier noise of the dissector, the high-light stability of the iconoscope, the sensitivity increment of the image iconoscope, and the linearity of response and substantial avoidance of spurious shading at low lights of the orthicon. The area under its performance curve* (Figure 6) is considerably larger than that under any of the other pickup tubes or under Super XX film but still considerably smaller than the area under the eye curve. Over a small range of its performance curve, its θ value from Eq. (8) is closely equal to the θ value for its photocathode and also to that for the eye. In other words, it successfully counts all absorbed quanta and achieves ideal operation. Towards low lights the counting process becomes obscured by proportionately more and more spurious counts from unused electrons in the scanning beam and the curve drops away from ideal performance. Towards high lights its signal remains constant with increasing light and again its performance curve departs from ideal performance. Another nonideal characteristic of the tube, and one that cannot be conveniently shown in Figure 4, is that only the high-light portions of a given picture are properly represented by this curve. The low-light portions of the picture fall along a line of about a 45° slope whose upper end is fixed by the curve in Figure 4. This is another way of stating that the absolute noise content of the low-light parts of a picture is the same as that for the high lights. For ideal operation the absolute noise content of each part of a picture should be a function of the brightness of that part and in particular should decrease with decreasing brightness.

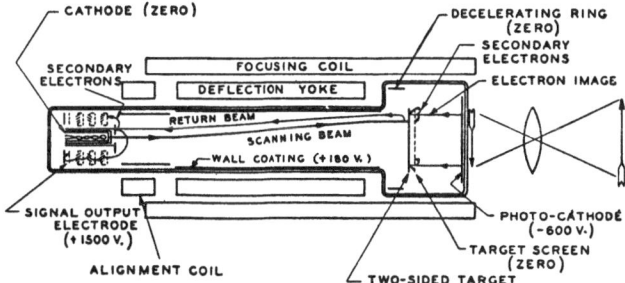

FIGURE 9. Image orthicon.

A rapid trip through the tube will perhaps clarify some of the above points. The optical image is focused on a semitransparent conducting photocathode whose sensitivity, in general, can be made higher than that

---

* For effective operation at low lights the target capacitance of the image orthicon should be reduced from its optimum value for high lights.

of an insulating target. The resulting photoelectrons are focused by a uniform magnetic field and accelerated to about 300 volts at which velocity they strike the back side of a two-sided target. Here, by virtue of a secondary emission ratio greater than unity, an amplified positive charge pattern is formed on the target. The secondary electrons are collected by a fine mesh screen located very close to the target. The potential of this collector screen also limits the potential to which the target can be charged by the picture. Stability at high lights is thereby achieved as well as the leveling off of the signal vs. light curve. The two-sided target is a thin sheet of glass. It is thin enough so that the scanning beam on the other side of the target can clearly see the picture charge and deposit, as in the orthicon, an equal amount of negative charge. The resistivity of the glass is adjusted so that these two charge patterns of opposite sign continuously unite and neutralize each other by conduction. The part of the scanning beam that is not deposited on the target returns to an electron multiplier located at and around the electron gun. The gain of the multiplier is high enough to raise the signal and noise level of the electron beam above the noise level of the amplifier to which the tube is connected. While the actual gain of the multiplier may be about five hundred, the useful gain (useful for sensitivity) varies from about twenty for high-light pictures to several hundred for low-light pictures.

From this description it can be seen that when all the photoelectrons are stored on the target (as positive charges by secondary emission) and when most of the beam electrons also land on the target to neutralize the photo charge, the tube can make a fairly accurate count of the number of absorbed quanta. Under these conditions its operation and performance are ideal. At very low lights, however, only a small fraction of the beam electrons, for electron optical reasons, are useful for counting charges on the target. The rest contribute a background noise that obscures the desired count. At high lights, not all the photoelectrons are stored on the target. They could be if the potential of the fine mesh screen collector were made arbitrarily high. But to preserve picture quality, this potential is kept at a reasonably small value limiting the amount of charge that can be stored.

To compute the image orthicon curve in Figure 4, a photocathode sensitivity of 10 microamperes/lumen was taken and a focal length of 3 in. Lens diameter and exposure time, as for the other devices, were set at 0.3 in. and 0.2 sec respectively.

4. Discussion of Performance Curves

The method of plotting and the curves in Figure 4 are not the most convenient for deciding what pickup device should be used for specified applications. Such would be the case if the performance curves for all of the devices were ideal and if there were no questions of ranges of contrast and angular size to be reproduced, spurious signals, freedom from distortion, stability, size, and so on. The departures from ideal performance, however, and the need for satisfying many side conditions make the choice of a device for a given purpose a matter for careful compromise.

Figure 4 was designed to emphasize these points:

1) The gap between the performance of actual pickup devices and maximum theoretical performance.

2) The relatively large range and high level of performance of the eye.

3) The relatively narrow ranges and generally low level of performance of pickup devices other than the eye.

4) The inadequacy of defining the sensitivity of nonideal devices by a single number. Such a number has meaning only for a specified scene brightness or at most a small range of brightnesses. In general, other qualifications also need to be stated.

5) The simplicity of the sensitivity scale for ideal devices. A single number, the quantum efficiency of the primary photoprocess, suffices (see dotted line marked $\theta = 0.1$).

The items just listed are easily assessed by inspection of Figure 4. Figure 4 does not, however, suggest the steady improvement of picture quality that accompanies increased scene brightness. If Figure 4 were rotated through 45° so that the line for $\theta = 1$ sloped upwards, the desired effect would be obtained. The same would result if $B\theta$ instead of θ, alone, were used to measure performance. In Figure 10, the curves of Figure 4 are replotted using $B\theta$ for the vertical axis. On this figure, the various levels of performance, $\theta = 1$, $\theta = 10^{-1}$, $\theta = 10^{-2}$, etc., would be marked out by lines parallel to the line marked $\theta = 1$ but shifted one order of magnitude to the right each time. Also on this figure, and this is its purpose, a horizontal line marks out a line of equal picture quality. By way of example, a horizontal line drawn at $B\theta = 10^{-2}$ intersects all of the curves at different scene brightnesses. Its meaning may be stated as follows: The picture that the eye sees at 1 ft.-L could be seen also by an ideal device with $\theta = 1$ at 10^{-2} ft.-L; by an image orthicon or by Super XX at 2 ft.-L, by an orthicon at 200 ft.-L, by an iconoscope at 600 ft.-L and by an image dissector at 10^5 ft.-L, all of these devices having the same exposure time and depth of focus as the eye. It is seen here immediately that if one had chosen to draw the horizontal line at another value of $B\theta$, the relative sensitivities of the various devices, owing to their departures from ideal performance, would not have been the same.

Consider another horizontal line drawn tangent to the highest point on the Super XX curve. This also intersects the eye curve at about 3 ft.-L and says that the picture quality seen by the eye at 3 ft.-L matches the best quality that Super XX can transmit. By quality here is meant half-tone discrimination or signal-to-noise ratio. If the eye, then, looks at motion pictures taken with 35-mm Super XX and if the picture subtends an angle of 25° at the eye, the eye will not be critical of the shortcomings of the film, providing the motion picture screen brightness is 3 ft.-L or less. If the screen brightness is raised, the discrimination of the eye exceeds that of the film and the eye becomes conscious of the noise or graininess of the film. Since motion picture screen brightnesses are nearer 10 than 3 ft.-L, finer grained films requiring higher scene brightnesses than Super XX are used for good quality pictures.

The same considerations hold in somewhat exaggerated form when, for example, the eye looks at kinescope pictures whose brightness may be around 50 ft.-L and the pictures are those transmitted by an image orthicon picking up a scene whose brightness may be only a few foot-lamberts. If the television camera were to match the depth of focus of the eye, the resultant kinescope pictures would be exceedingly noisy. The eye would in a sense be looking at its own ft.-L quality pictures with a sense of discrimination corresponding to tens of foot-lamberts. If the eye could accomplish this feat more directly it would indeed be critical of its own low light performance. Presumably, the brightness or gain controls in the eye are more nicely matched to the eye's performance than are the arbitrarily variable knobs on a television receiver.

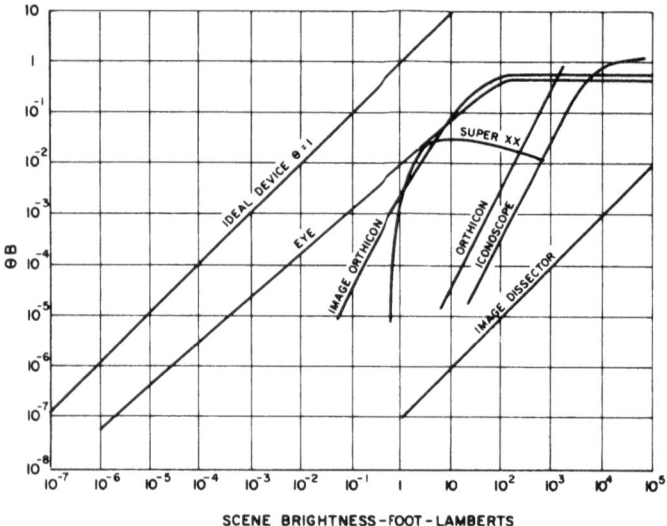

FIGURE 10. Performance curves of Figure 4 replotted with Bθ as ordinate.

The conditions outlined above — namely, observer looking at a bright kinescope and camera looking at a dim scene — can still result in noise-free pictures if the camera lens is opened up at the expense of depth of focus. This is generally a quite acceptable compromise. The alternative solution, a more sensitive camera, hangs chiefly on the development of more sensitive photocathodes. And all too little is known of the workings of present photosurfaces, let alone improved ones.

The problems of noise visibility just discussed arise frequently and in varied form. They can lead to particularly complex analyses when many parts of a system must be considered. A method of analysis that has wide applicability and that avoids these complications will be outlined in a following section. Another sensitivity problem will be discussed here first.

In a recent paper [19] the writer evaluated a figure of merit designed to give the relative sensitivities of photographic film and the human eye. The ratio was about 200:1 in favor of the eye and was based essentially on the facts that the scene brightness and lens area for the motion picture camera that records pictures are each about ten times the screen brightness and lens area for the observer who views them. This ratio (200) would appear to be inconsistent with the curves in Figure 10 which show film approaching closely the performance of the eye at least near the low-light end of the film curve. A major part of the discrepancy, however, appears to be removable if due account is taken of the relatively sharp threshold for film. For example, if an observer and a motion picture camera both look at a scene whose high lights lie at the point where the curves for eye and film in Figure 10 are closely matched, these high lights will be reproduced approximately with equal quality. But the low lights of the scene will appear black on the film by virtue of its sharp cut off, while for the eye they will appear with many shades of grey. In fact, the eye very seldom sees true blacks in a scene. If the range from high lights to low lights in the scene is of the order of a hundred fold, the camera would need the same order of increase of scene brightness or exposure in order to avoid rendering black what the eye normally sees as grey. There are other considerations, such as the gamma of the final photographic print, that influence the relative amounts of light required in taking and in viewing motion pictures, but the above argument would appear to offer the largest factor.

VIII. A CRITERION FOR NOISE VISIBILITY

Noise in a television picture is common enough to be familiar to almost any one who has looked at television pictures. The counterpart of noise in a television picture is graininess in photographic film. But, to the casual observer of motion pictures, graininess is neither frequent nor prominent. That is mainly because film is viewed under conditions more carefully controlled than they are for television pictures. Increase the brightness of the motion picture screen, or let the observer come arbitrarily close to the screen, or use the "fastest" film material and motion picture graininess would be as common as television noise. Whether or not the usual observer is aware of graininess in the pictures he witnesses, is not of primary importance in this discussion. What is to the point is that the existence of graininess has already imposed its restrictions on the brightness of motion picture screens, the minimum approach to such screens, and the scenes that one chooses to photograph. Again, it would be even more difficult to find observers who agreed that they were annoyed by fluctuations in their own visual processes. Yet there is good reason to believe that such fluctuations do limit what can be seen. In brief, the performances of the several devices — television pickup tubes, photographic film, and the eye — are circumscribed by noise, more or less prominently displayed. When two or more of these devices are combined into a system, it is desirable to be able to assess the relative noise contributions of each.

But, in a system involving a scene of variable brightness, a camera of specified performance and adjustable optical geometry, a receiving screen of variable size and brightness and an observer at an arbitrary viewing distance, the problem of locating the particular noise source that constitutes the bottle-neck for performance is not, in general, a simple one. For many problems of this type, however, a method of analysis which is at least conceptually simple, and usually operationally so, will be described. It is especially applicable to systems whose components show ideal performance.

The principle of the method may be outlined by considering a multistage electron multiplier. The signal-to-noise ratio coming out of such a multiplier is very closely the signal-to-noise ratio of the electron stream entering the multiplier. For any n o r m a l multiplier this is true. The noise contribution of the separate stages, especially for high gains per stage, is negligible. Suppose, however, that a stage somewhere in the middle of the multiplier turned out to have a gain less than unity. The signal-to-noise ratio out of the multiplier would still not be affected providing that the product of the gain of this stage and the total gain of all preceding stages was greater than unity. If this product were less than unity, the output current of the low gain stage would then determine the signal-to-noise ratio out of the multiplier. Or, generally, if one measured the electron currents coming into each stage of the multiplier the lowest of these currents would determine the final signal-to-noise ratio.

Similarly, the many transformations through which a picture element of the original scene goes before it emerges as a picture element in the brain of the final observer may be regarded as multiplierlike stages having various gains. The problem of finding the noise bottleneck in such a system becomes one of finding the smallest stream of particles or events representing the original picture element. An illustration of the method of analysis follows.

Figure 11 shows schematically the elements of a system from scene through pickup tube and receiver to the final observer. Nine points of interest are marked out. At each point, the number of "events" representing an element of the original scene is indicated algebraically. At the bottom of the figure the numbers are arranged in a qualitative plot. Starting from an element of the original scene, N_o quanta are emitted per second in a lambert distribution. $N_o \sin^2 \phi_1$ of these quanta pass through the pickup tube lens whose half angle subtense at the scene is ϕ_1. At the photo surface of an ideal pickup tube, the $N_o \sin^2 \phi_1$ quanta are converted into a still fewer number $(\theta_1 N_o \sin^2 \phi_1)$ of photoelectrons, θ_1 being the quantum yield of the surface. The pickup tube multiplies these signal electrons by the factor G_1, which is large enough that noise in the circuits connecting pickup tube and receiving tube may be neglected. These circuits further amplify the signal current by the factor G_2 so that the number of electrons in the receiving tube that represent the original element is about a thousand times greater than the number leaving the photocathode of the pickup tube. All of the gain factors are assumed to

be constants characteristic of a l i n e a r * system. The electrons in the
receiving tube strike a luminescent screen and are converted into about
300 times as many light quanta. G_3 is this factor. If the quanta from the
luminescent screen also follow a lambert distribution, the number passing
through the eye lens is a small fraction, $\sin^2\phi_2$, of the total number. Only
a fraction θ_2 of those entering the eye are usefully absorbed at the retina.
θ_2 is the quantum yield of the retina. Finally, some sort of gain element
must amplify the effect of these absorbed quanta before they can generate
the nerve pulses carrying information to the brain.

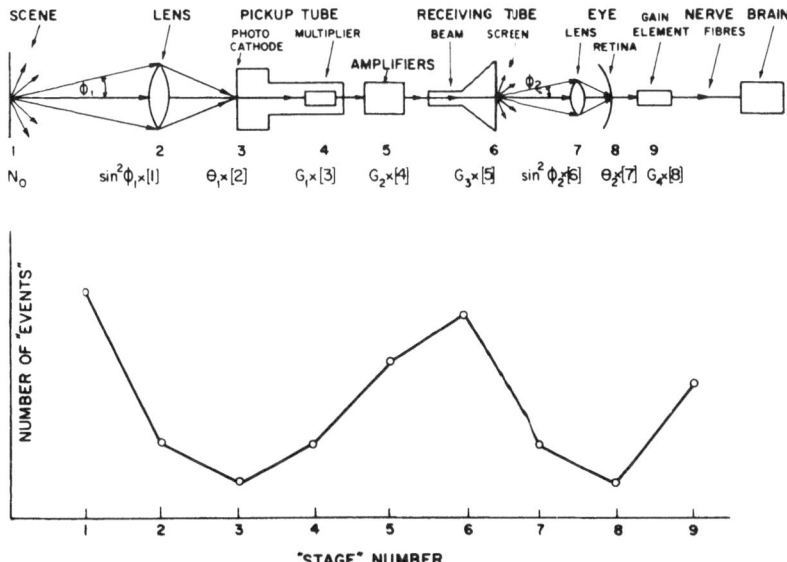

FIGURE 11. Chronicle of a picture element from the original scene to the brain of
the final viewer.

It is seen from the plot in Figure 11 that there are two low points, one
at the photosurface of the pickup tube and the other at the photosurface of
the eye. If the pickup tube point is lower than that of the eye, the
observer sees pickup tube noise on his receiving tube screen and the
system performance is limited by the pickup tube. For the reverse order,
the received picture is judged noise free and the system performance is
limited by the eye. Technical economy would suggest a close match
between limitation by eye and tube.

---

* For a nonlinear system, the v i s i b i l i t y of noise on the receiving tube screen is proportional to the gamma
of the system. The gamma of the system does n o t, however, vary the s i g n a l – t o – n o i s e r a t i o and,
therefore, the intelligence transmitted. These statements are not to be confused with the arrangement of
low gamma transmitter and high gamma receiver, which is designed to minimize the effects of noise picked
up in transmission. The present discussion assumes that such noise may be made negligible.

135

Figure 11 shows also how the transition from a noisy to a noise-free picture can be made simply by decreasing the screen brightness at the receiving tube or by increasing the viewing distance. For the particular case in which the two brightnesses, angular subtenses and quantum yields are made equal, the pickup tube becomes in fact a transposed eye.

IX. INTELLIGENCE VS. BANDWIDTH AND SIGNAL-TO-NOISE RATIO

In the introduction to this paper, the phrase "bits of information" was used to characterize the intelligence transmitted by a pickup device. This usage was not by way of popularizing otherwise technical verbiage but rather to emphasize the finite character of the intelligence. That is, only a finite number of spatially separable elements and only a finite number of half tone steps per element are recognizable in a picture having a specified signal-to-noise ratio. In fact, one can easily count the total number of possible pictures that can be compounded out of n separate elements and m half-tone steps per element. That number is m^n. On the other hand, the capacity of a system, having a bandwidth Δf and a signal-to-noise ratio R, for transmitting intelligence may also be measured by the total number of different pictures it can transmit. This number by the same reasoning is $\left(\dfrac{R}{k}\right)^{2T\Delta f}$ where k is the threshold signal-to-noise ratio and T is the time for one picture. Thus for each of the $2T\Delta f$ separable elements of time assigned to a picture there are R/k distinguishable values of signal amplitude that may be selected. Let the desired number of pictures be set equal to the number of different pictures that the system can transmit:

$$m^n = \left(\frac{R}{k}\right)^{2T\Delta f} \tag{9}$$

The solution of this equation for Δf yields:

$$\Delta f = \frac{n}{2T}\,\frac{\log m}{\log\left(\dfrac{R}{k}\right)} \tag{10}$$

Eq. (10) says that when the signal-to-noise ratio is just large enough for the discrimination of the desired number of half tone steps, namely, if $\dfrac{R}{k} = m$, the bandwidth has its customary value of half the number of picture elements per second. If, on the other hand, the signal-to-noise ratio is larger than it need be for half tone discrimination, the bandwidth may be reduced by the factor $\dfrac{\log m}{\log R/k}$.* A specifically designed mechanism is needed, however, to effect this reduction.

A final comment should be made concerning bandwidths in excess of that needed to transmit the intelligence content of a picture. An over-

---

* This reciprocity of bandwidth and signal-to-noise ratio (or its logarithm) was pointed out to the writer several years ago (1944) by Dr. G. A. Morton of RCA Laboratories Division.

size bandwidth does no particular harm to the picture transmitted by an ideal device. The noise content of such a picture is set by the picture itself and not by the bandwidth of the associated circuits. An exaggerated example of this is the series of photographs in Plate III. Here the bandwidth was about 5 "megacycles" while the picture content, for some of the shorter exposures, needed only a few kilocycles to convey substantially all of its information. Of course, wider bandwidths make it more difficult to transmit a picture without picking up noise in the transmission comparable with noise in the original picture.

Another example, effectively, of extremely wide band transmission is found in the photographic process. Using a microscope one can see the separate photographic grains whose size is of the order of a micron. Under normal conditions of viewing, however, the eye can see only elements larger than about 25 microns. That is, the eye sets the effective bandwidth at less than 1% of the capacity of the film. In fact, it is of no great consequence whether the eye can see the limiting resolution of film or not. The intelligence contained in elements near the limiting resolution of film is relatively small by virtue of the inherently low signal-to-noise ratio of these elements. For this reason it is not sensible to compare without qualification the number of lines of a television picture, for which number of lines the signal-to-noise ratio may be as high as needed, with the limiting resolution of film where, by definition, the signal-to-noise ratio approaches zero. A valid comparison must depend on relatively subjective tests which in turn are a critical function of picture content. Such tests and analyses have been carried out by Baldwin [21] and, more recently, in an extensive and thorough investigation by Schade [4]. Schade has found, for example, that equal picture quality can be transmitted by a television system having a number of scanning lines equal to half the limiting resolution of a grainy film.

Oversize bandwidths, while not penalizing the performance of an ideal device, do deteriorate the rendition of blacks and greys by a device, like the orthicon or iconoscope, for which amplifier noise is the limiting noise source.

X. CONCLUDING REMARKS

Lest the original purpose of this discussion not be recognizable in its elaboration, it is here restated. Because the final limitations of pickup devices are clearly and simply set by the quantum nature of light, because these limitations have not been widely discussed, and because the particular devices in the new and rapidly developing field of television are of less interest for the mechanics of their operations than they are as markers in the approach to ideal performance, the emphasis throughout has been on the setting up of an absolute scale of performance according to which the many and diverse pickup devices can be oriented. This approach is particularly needed because the eye, photographic film, and television pickup tubes are being called upon more and more frequently to critically pass upon each other's performance.

REFERENCES

1. G. C. SZIKLAI, R. C. BALLARD and A. C. SCHROEDER, Proc. Inst. Radio Engrs., **35**, 862 (1947)
2. A. ROSE, Proc. Inst. Radio Engrs.. **30**, 295 (1942)
3. H. DeVRIES, Physica, **10**, 553 (1943)
4. O. H. SCHADE, RCA Rev., **9**, 5 (1948)
5. P. W. COBB and F. K. MOSS, J. Franklin Inst., **205**, 831 (1928)
6. J. P. CONNOR and R. E. GANOUNG, J. opt. Soc. Am., **25**, 287 (1935)
7. H. R. BLACKWELL, J. opt. Soc. Am., **36**, 624 (1946)
8. P. REEVES, Psychol. Rev., **205**, 831 (1928)
9. A. ROSE, J. opt. Sci. Am., **38**, 196 (1948)
10. S. HECHT, J. opt. Soc. Am., **32**, 42 (1942)
11. L. A. JONES and G. C. HIGGINS, J. opt. Soc. Am., **36**, 203 (1946)
12. V. K. ZWORYKIN and G. A. MORTON, Television, Wiley, New York, 1940
13. V. K. ZWORYKIN and E. G. RAMBERG. Section 15, part II, Electrical Engineer's Handbook, Ed. by H. Pender and K. McIlwain, 4th ed. Wiley, New York, in press.
14. P. T. FARNSWORTH, J. Franklin Inst., **218**, 411 (1934)
15. C. C. LARSON and B. C. GARDNER, Electronics, **12**, 24 (Oct., 1939)
16. V. K. ZWORYKIN, G. A. MORTON and L. E. FLORY, Proc. Inst. Radio Engrs., **25**, 1071 (1937)
17. H. A. IAMS, G. A. MORTON and V. K. ZWORYKIN, Proc. Inst. Radio Engrs., **27**, 541 (Sept., 1939)
18. A. ROSE and H. A. IAMS, RCA Rev., **4**, 186 (1939)
19. A. ROSE, P. K. WEIMER and H. B. LAW, Proc. Inst. Radio Engrs., **34**, 424 (1946)
20. A. ROSE, J. Soc. Motion Picture Engrs., **47**, 273 (1946)
21. M. W. BALDWIN, Proc. Inst. Radio Engrs., **28**, 458 (1940)

IV. An Analysis of the Gain–Bandwidth Limitations of Solid–State Triodes

I. INTRODUCTION

The attempts to extend the performance of triodes to higher frequencies (e.g., above one gigacycle) have led various individuals to stress the virtues of one or more of the triode arrangements sketched in Figs. 2-6. Johnson and Rose [1] have singled out the unique property of the bipolar transistor in that it achieves the closest physical coupling between the controlling and the controlled charge. Wright [2] has emphasized the short transit times attainable in the space-charge-limited triode [3]. The author had proposed a majority-carrier triode (referred to at the time as the "Barristor") using a thin electron-permeable metal grid to avoid the limitations of the base resistance of the bipolar transistor. Mead [4] has proposed a "tunnel triode" in which the high emitter current densities are expected to lead to improved performance at high frequencies. Finally, the well-known unipolar, or field-effect, transistor [5,6] has the advantage of being a simple, majority-carrier device.

It is the purpose of this analysis to show that none of the solid-state triodes has a unique advantage in its principle of operation at high frequencies, and that the upper limit to the high-frequency performance of all of the listed triodes is controlled in much the same way by the same physical parameters, of which the major parameter is the dielectric relaxation time of the space between emitter and grid. This conclusion does not preclude an emphasis on particular devices for particular application for technological reasons, especially at frequencies below the upper limit.

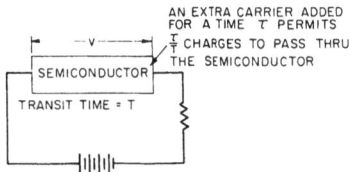

FIGURE 1. Generalized argument for current gain.

Reprinted from RCA Rev., 24 (4), 627-646 (1963).

This structure was described in a Government Contract report (September 1960) on work supported in part by the U. S. Navy. It has since been described independently by M. M. Atalla at the IRE-AIEE Device Research Conference, Durham, N. H., 1962, and by D. V. Geppert, 'A Metal-Base Transistor,' Proc. Inst. Radio Engrs., 50, 1527 1962.

II. GAIN-BANDWIDTH PRODUCT

We review quickly the generalized formulation [1] of the gain-band-width product for charge-control devices.

Consider, as in Figure 1, two electrodes connected ohmically to a semiconductor. If we add an extra carrier to the semiconductor for a time τ, the extra carrier will allow τ/T extra electron charges to pass between the electrodes, where T is the transit time of a carrier between electrodes. The purpose of any control grid is to add (or subtract) extra carriers in the region between emitter and collector. If a charge $+Q$ is put on the grid, an equal and opposite charge $-Q$ is introduced into the emitter-collector space. The grid or control current is then Q/τ, and the collector or controlled current is Q/T. The current-gain is then:

$$\text{Current gain} \equiv G_I = \frac{\tau}{T}.$$ (1)

Now, to put the charge Q on the grid requires a voltage

$$V_i = \frac{Q}{C_i}$$

where C_i is the input or grid capacitance. The charge on the grid gives rise to the amplified charge $Q\tau/T$ at the collector. The voltage appearing at the collector will then be

$$V_o = \frac{Q}{C_o} \frac{\tau}{T}$$

where C_c is the output or collector capacitance; the voltage gain will be

$$G_V = \frac{C_i}{C_o} \frac{\tau}{T} = \frac{C_i}{C_o} G_I.$$ (2)

The power gain of the device can now be written as

$$G_P = G_I G_V = \frac{C_i}{C_o} \left(\frac{\tau}{T}\right)^2.$$ (3)

The bandwidth of the device is given by

$$\Delta f = \frac{1}{2\pi\tau}$$ (4)

From Eqs. (3) and (4).

$$(G_P)^{1/2}\Delta f = \left(\frac{C_i}{C_o}\right)^{1/2} \frac{1}{2\pi T}.$$ (5)

140

Note that in this argument it is assumed that the total input capacitance is between the controlling charge on the grid and the controlled charge introduced into the emitter-collector space. Other sources of input capacitance detract from the performance of the device. Also, it is assumed that the resistance of the control grid itself is negligible. This is not true in present high-frequency bipolar transistors. Their base resistance is a significant limitation to high-frequency performance. Finally, the collector is assumed to have negligible d-c leakage to other electrodes.

III. DISCUSSION OF GAIN-BANDWIDTH PRODUCT

Inspection of Eq. (5) shows that a large ratio of C_i/C_o is advantageous. In most triodes C_i and C_o are geometrical capacitances determined by electrode dimensions and separations. One would not expect any unusual differences to exist in these parameters for the several types of triodes, provided they have comparable dimensions. The bipolar transistor is the notable exception in that C_i is the sum of a geometrical capacitance (mainly base to emitter) and a diffusion capacitance that ordinarily is many times larger. In fact, the diffusion capacitance represents the high-capacitance limit when controlling and controlled charge are made to approach each other and to occupy finally the same physical space.

These considerations led Johnson and Rose [1] to single out the bipolar transistor as having a special advantage over other triodes. The argument, however, is not complete for the following reason. C_i is proportional to the base thickness, d. The transit time (through the base) is proportional to d^2. Consequently if one reduces the base thickness, more is gained from the shorter transit time than is lost by the smaller value of C_i. In fact it is profitable to reduce the base width at least until the transit time across the base is as small as the transit time between emitter and base or base and collector. We show below that, under these conditions, the diffusion capacitance becomes equal to the geometric capacitance between base and emitter, the geometric capacitance being that across a Debye length. The meaning of this result is that even though the controlling and the controlled charges (electrons and holes) occupy the same physical space in the base, the effective geometrical spacing to be associated with the diffusion capacitance is a Debye length of the minority carriers. Hence the diffusion capacitance can be matched in a unipolar device such as the field effect transistor by scaling the critical dimensions down to a Debye length. The net result is that the high input capacitance of the bipolar transistor does not lead to higher gain-bandwidth products than can be achieved by other triodes.

It remains then to examine the transit time, T, for the separate devices. It will be shown that this transit time can be traced in all cases to an RC time constant, a transit time across a depletion layer, or a transit time across a space-charge-limited current element. Further, it will be shown that these three times are identical (assuming the drift velocity of carriers does not saturate) for materials having the same conductivity and dielectric constant — that is, for materials having the same dielectric relaxation time $(\tau_{rel} = (K/(4\pi\sigma)) \times 10^{-12} \text{ sec})$.

141

The conclusion then follows that the upper limit of frequency response of all of the triode devices is the dielectric relaxation time of the material as used in the device. It goes almost without saying that lower performance can result from improper design or operation. Further, the ease of realizing the maximum performance may differ among the several devices for significant technological reasons.

By way of calibration, the dielectric relaxation time of one-ohm-cm material is about 10^{-12} sec.

IV. ANALYSIS

It is first shown that $T = RC$ for both a depletion layer and for a space-charge-limited current element.

The thickness of a depletion layer is given by

$$Q = CV \tag{6}$$

or

$$ned = \frac{K}{4\pi d} V \times 10^{-12}, \tag{7}$$

where Q is the charge per cm^2 in the depletion layer, C its capacitance, and V the voltage step across the layer. V is the sum of the contact potential and the applied voltage, d is the thickness of depletion layer, K is the relative dielectric constant, and n is the volume density of "available" charge. "Available" means free to move within the time of the experiment, and includes both free and trapped carriers. From Eq. (7),

$$d^2 = \frac{K}{4\pi ne} V \times 10^{-12}, \tag{8}$$

and

$$T = \frac{d^2}{V\mu} = \frac{K}{4\pi ne\mu} \times 10^{-12} \text{ sec.} \tag{9}$$

Also, for a leaky capacitor it is known that

$$RC = \frac{K}{4\pi\sigma} \times 10^{-12} = \frac{K}{4\pi ne\mu} \times 10^{-12} \text{ sec.} \tag{10}$$

From Eqs. (9) and (10) it is concluded that $T = RC$. The error in regarding the depletion layer as a simple, rather than distributed, capacitor is less than a factor of 2, and enters in a self-cancelling manner in the computation of both T and RC.

Next it is shown that $T = RC$ for a space-charge-limited current element. The free charge introduced by the applied voltage in a space-charge-limited diode is

$$Q = CV. \tag{11}$$

142

The corresponding free carrier density is given by

$$Q = ned = \frac{K}{4\pi d} V \times 10^{-12}. \tag{12}$$

From Eq. (12),

$$T = \frac{d^2}{V\mu} = \frac{K}{4\pi ne\mu} \times 10^{-12} = RC. \tag{13}$$

The transit times for the several triodes will now be examined individually beginning with the bipolar transistor.

Bipolar Transistor (Figure 2)

For simplicity we begin with a bipolar transistor in which the thickness of base has negligible effect on the transit time of carriers from emitter to collector. The implications of this assumption will be discussed later. Also, let the recombination lifetime for carriers in the base be large relative to the reciprocal of the frequency of operation.

If a positive charge is now added to the base (emitter is grounded), negative charge will be drawn in from the emitter to neutralize it. The time required for this neutralization is the RC time constant of the emitter-base electrodes considered as a simple capacitor. Since the negative charge drawn in from the emitter does not remain in the base, but is passed on to the collector, the ratio of total charge passed to the charge on the base is

$$\frac{Q_{coll}}{Q_{base}} = \frac{\tau}{RC}, \tag{14}$$

where $\tau = 1/(2\pi f)$ and f is the operating frequency. Eq. (14) also represents the current gain because both Q_{coll} and Q_{base} are referred to the same element of time, τ.

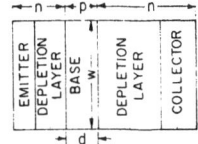

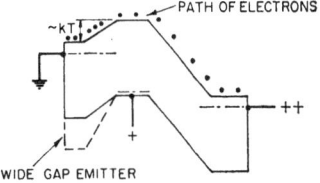

FIGURE 2. Bipolar transistor. The relative dimensions have been chosen to fit the preferred arrangement discussed in the text. The wide-gap emitter needed to permit a highly conducting base is shown in dashed outline.

The resistance R between emitter and base is localized mainly in the last kT rise in potential in the depletion layer bordering the base. This resistance is reduced exponentially since the emitter is forward biased. Hence, it is of advantage to operate with the conduction band of the emitter within about kT of the conduction band of the base. Under these conditions the RC of the material lying between emitter and base is given closely by the dielectric relaxation time of the emitter material, namely,

$$RC = \frac{K}{4\pi\sigma_{emitter}}.\tag{15}$$

From Eqs. (14) and (15) it is seen that the characteristic transit time that is to be inserted into Eq. (5) (the general relation for gain-bandwidth product) is

$$T = RC = \frac{K}{4\pi\sigma_{emitter}}.\tag{16}$$

In arriving at Eq. (16) it has been assumed that the transit time across the base is less than the time cited in Eq. (16) for transit of charge from emitter to base. Since, by Eq. (9), the RC time of Eq. (16) is also the transit time of carriers through the depletion layer when a potential of kT/e is applied, and since the effective potential across the base layer for the transit of carriers through the base is also kT/e, it follows that the base thickness should be less than the depletion layer thickness in order that the transit time through the base be less than the RC of the emitter-base depletion layer.

The same criterion, base thickness less than depletion layer thickness, is also the criterion for ensuring that the diffusion capacitance of the base is less than the geometric capacitance between emitter and base. This can be seen from the following two relations:

$$\text{diffusion capacitance} = \frac{(ned)_{base}}{kT/e}\tag{17}$$

geometric capacitance (from Eq. (6) and (7))

$$= \frac{(ned)_{depletion}}{kT/e}.\tag{18}$$

The free-carrier densities n in Eqs. (17) and (18) are very nearly the same since the emitter and base conduction bands are within kT of each other.

In order that the high-frequency performance given by Eq. (16) be realized it is, of course, important that there be no other time-limiting processes. If the collector is made of material at least as conducting as the emitter, then by Eq. (9) the transit time from base to collector will be at least as short as that from emitter to base. This statement is based on a nonsaturated drift velocity. If the drift velocity is saturated, the transit time in the collector region can exceed that in the emitter region. The series resistance of emitter and collector leads will not impose limitations provided the thickness of emitter and collector layers in series with their respective depletion layers is less than that of the depletion layers. (Alternatively, of course, these series layers could be designed to be thicker but proportionately more conducting, i. e., by diffusion or expitaxial growth.) The series resistance to the base layer, namely, the sheet resistance of the base, is a more serious problem. Here the conductivity of the base must exceed the conductivity of the emitter by the product of their aspect ratios, that is, the ratio of width to thickness. But this is highly unfavorable for emitter efficiency. The latter requires that the conductivity of the base be actually less than that of the emitter. One can still think in terms of a highly conducting base without adverse effect on emitter efficiency provided the forbidden gap of the emitter is appreciably larger than that of the base. Once one admits the high-band-gap emitter into the design possibilities of a bipolar transistor, one has only a short step to make to arrive at the Barristor — an essentially majority-carrier device.

In summary, one can achieve the high-frequency limit of a bipolar transistor given by the RC of its emitter provided a high-band-gap emitter is used and provided the thickness of emitter and base materials is not more than twice the thickness of their respective depletion layers.

Barristor (Figure 3)

It is now relatively easy to discuss the Barristor in terms of the arguments already detailed for the bipolar transistor. The characteristic time governing the high-frequency performance is again the RC time constant between emitter and grid.

No other limiting times enter in, since the same arguments used for the bipolar transistor hold here. The thickness of emitter and collector layers should be no greater than twice the thickness of their respective depletion layers. The conductivity of the collector should be at least as large as that of the emitter. The transit time of carriers through the metal grid will be negligible since the thickness of metal grid should be less than a few tens of angstroms to permit a high transparency for carriers going from emitter to collector. Since the metal grid is so much thinner than the emitter depletion layer, its conductivity must be correspondingly higher. In fact, to avoid limitation by the sheet resistance of the metal grid, the conductivity of the grid must be as large as

$$\sigma_{grid} = \sigma_{emitter} \frac{W^2}{d_{grid}\, d_{depletion\ layer}} ;$$

W is the sheet dimension of the grid and d_{grid} its thickness.

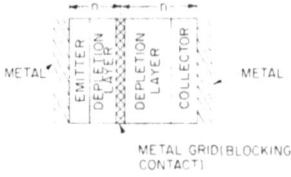

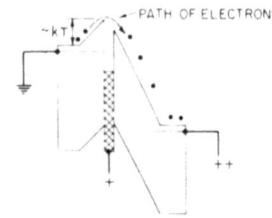

FIGURE 3. Barristor. The difference in barrier heights on the two sides of the metal grid is obtained either by using different semiconductors for emitter and collector or by using different metals on the two faces of the grid.

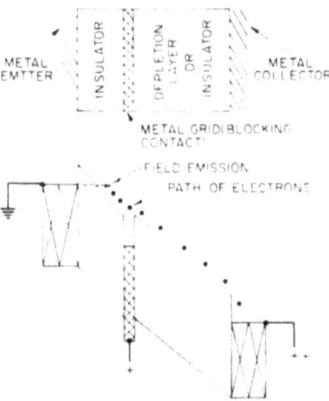

FIGURE 4. Field-emission triode. While Mead's proposal used an insulator on the collector side of the grid, a depletion layer can be used as well.

Field-Emission or Tunnel Triode (Figure 4)

It is clear from Figs. 3 and 4 that the major difference between the field-emission triode and the Barristor lies in the means of introducing carriers into the emitter-grid space. The field-emission triode does so by causing carriers to tunnel from the emitter into the insulator separating emitter from grid. The Barristor does so by the well-known forward bias characteristics of a metal-semiconductor rectifier. While Mead emphasizes the high current density available by field emission from a metal, this advantage is tempered by the difficulty of attaining uniform, nonerratic field emission, and by the questionable need for such high current densities. For example, the Barristor can already, in principle, achieve response times of 10^{-13} sec by using 10^{-1} ohm-cm material for its emitter, which implies current densities of the order of 10^4 amperes/cm². It remains to be seen whether such short times are even useable in a triode amplifier.

In brief, the field-emission triode is limited by the same RC time constant in the emitter-grid space as the Barristor and bipolar transistor. The side conditions on the emitter and collector leads as well as on the conductivity of the grid are the same as those for the Barristor.

Space-Charge-Limited Triode (Figure 5)

The space-charge-limited (SCL) triode begins with an insulating material to which electrodes are attached. The space between emitter and grid must be made conducting by the injection of carriers from the emitter. Under

146

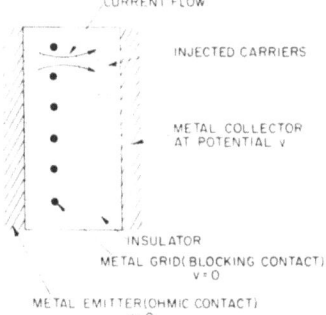

FIGURE 5. Space-charge-limited triode. The leakage of the collector field through the grid wires provides the accelerating field at the cathode necessary to cause injection of carriers.

conditions of space-charge-limited flow, the transit time between emitter and grid becomes equal (see Eq. (13)) to the RC time constant (dielectric relaxation time) of the space between emitter and grid. As in the previous devices, the conductivity of this space becomes the governing parameter for high-frequency performance. It is worth noting from Eqs. (9) and (13) that the same magnitude of voltage is required to either inject or deplete a given space-charge density of carriers in a given space.

Unipolar Transistor (Figure 6)

The unipolar transistor shown in Figure 6 differs from the usual geometry [5] in that the extent of grid along the direction of current flow is small compared with the separation of grid wires, and also in that there are many grid wires rather than the usual two. The latter extension serves to define a grid plane as in the other devices rather than just a grid line. The use of grid wires rather than long grid plates is not likely to alter the essential physics of the device.

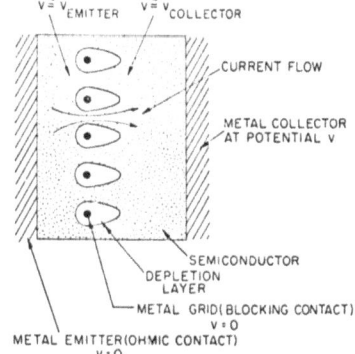

FIGURE 6. Unipolar transistor. By virtue of the geometry of current flow, most of the applied voltage lies across the narrow channels.

While a quantitative solution for the current flow in the case of grid wires has not yet been carried out, certain significant features of the solution can be estimated with reasonable confidence. Consider the case of grid and emitter grounded and the collector potential high enough to "saturate" the current flowing through the grid plane. Under these conditions

1) The length of constricted channel through which the current flows is approximately the long dimension of the depletion layer around the grid wires.

2) The potential drop along the channel is approximately the potential drop across the depletion layer. It is also approximately the potential applied to the collector.

From these statements and Eqs. (9) and (13), on can argue that the potential drop along the channel is almost sufficient to initiate space-charge-limited current flow. (It is not likely that in this geometry the density of carriers in the channel can at all be increased above their thermal equilibrium value by an applied voltage, since to do so would normally require a source of carriers close to the channel with an excess density of carriers.) We conclude then that the transit time of carriers from the emitter space to the collector space is given by the dielectric relaxation time of the semiconducting material as in the other four triodes discussed.

V. GENERAL REMARKS

It is, of course, not unexpected that the dielectric relaxation time of the space between cathode and grid should be the controlling parameter for high-frequency performance. The same conclusion couched in terms of conductivity or current density is well known for vacuum triodes. Also, the current-gain-bandwidth product for trap-free photoconductors [7] has been shown to be equal to the (dielectric relaxation time)$^{-1}$ of the photoconductor. The dielectric relaxation time is, by definition, the time required for a perturbation in charge or electric field in a medium to decay. It should measure the upper limit to the speed at which the medium can respond to changes in charge or field.

The current-gain-bandwidth product of trap-dominated photoconductors was obtained in the form

$$G_I \frac{1}{\tau} = \frac{1}{\tau_{rel}} M,$$

where τ is the response time to varying light signals, and τ_{rel} is the dielectric relaxation time. M is a factor dependent upon the energy distribution of traps and recombination centers. Its value is usually unity but it may greatly exceed unity for special trap distributions. The last statement may appear to contradict the expectation that the dielectric relaxation time is the fastest response time of a medium. The apparent contradiction is resolved by recognizing that it is the conductivity of the emitter-grid space that is significant. By contrast, the grid-collector space may be made considerably less conducting than emitter-grid space by increasing the collector voltage. This can be seen by inspection of the bipolar transistor or Barristor figures. When the bipolar transistor or Barristor is used as a photoconductor by allowing the grid to float, the distinction between these two spaces is more clearly resolved, and M becomes unity for the emitter-grid space and greater than unity if the grid-collector space is included [8]. In the case of the photoconductor the two spaces are not clearly separated. The emitter-grid space may be regarded as the space between emitter and virtual cathode. The rest of the space — the bulk of the photoconductor — may be regarded as the grid-collector space. The somewhat confusing feature is that the grid controlling charge is generated by optical excitation throughout the bulk of the photoconductor, that is, in the grid-collector space.

At first glance there appear to be large differences between the several triodes sketched in Figs. 2-6. One can de-emphasize these differences by starting with the Barristor, for example, and noting the small changes required to convert it into any of the other devices. If the metal grid is replaced by a degenerate p-type semiconductor, the Barristor becomes essentially the wide-gap-emitter bipolar transistor. If the metal is apertured instead of continuous, the barristor becomes a form of unipolar transistor. If, together with the apertured grid, the semiconductors are replaced by insulators, the Barristor becomes a space-charge-limited triode. The kinship of Barristor and field-emission triode has already been noted and is apparent by inspection.

VI. ACKNOWLEDGMENT

The writer is indebted to H. S. Sommers, Jr. and P. K. Weimer for critical readings of this manuscript.

REFERENCES

1. E. C. JOHNSON and A. ROSE, Simple general analysis of amplifier devices with emitter, control and collector functions, Proc. Inst. Radio Engrs., **47**, 407 (1959)
2. G. T. WRIGHT, A proposed space—charge—limited dielectric triode, J. Brit. I. R. E., **20**, 337 (1960)
3. W. RUPPEL and R. W. SMITH, CdS analog diode with triode, RCA Rev., **20**, 702 (1959)
4. C. A. MEAD, The tunnel—emission amplifier, Proc. Inst. Radio Engrs., **48**, 359 (1960)
5. W. SHOCKLEY, A unipolar "field effect" transistor, Proc. Inst. Radio Engrs., **40**, 1365 (1952)
   O. HEIL, British patent 439, 457 (1935)
6. P. K. WEIMER, The TFT — a new thin film transistor, Proc. Inst. Radio Engrs., **50**. 1462 (1962)
7. A. ROSE and M. A. LAMPERT, Gain—bandwidth product for photoconductors, RCA Rev., **20**, 57 (1959)
8. A. ROSE, Maximum performance of photoconductors, Helv. Phys. Acta, **30**, 247 (1957)

V. Photoconductive Photon Counters

The purpose of this discussion is to derive the conditions under which individual photons can be counted by means of a photoconductor plus amplifier. We will make use of the gain-bandwidth relation obtained in the previous lecture for photoconductors.

While the argument is carried out for a single photo element, its conclusions are valid for an array of such elements the array then constituting a system for recording images. The solid-state self-scanning sensor described by Weimer is a significant example of such an array.

If the elements of the array can detect individual photons, the system can then operate at arbitrarily low light intensities. If, on the other hand, n photons per element are needed to generate a signal above the noise, the threshold light intensity jumps from zero to n/a where a is the area of an element. A typical example is photographic film for which about 100 photons per grain are needed. For grains of micron size, this leads to threshold light intensities on the film (0.½/second exposure) of about 10^{-2} ft.-L.

CIRCUIT

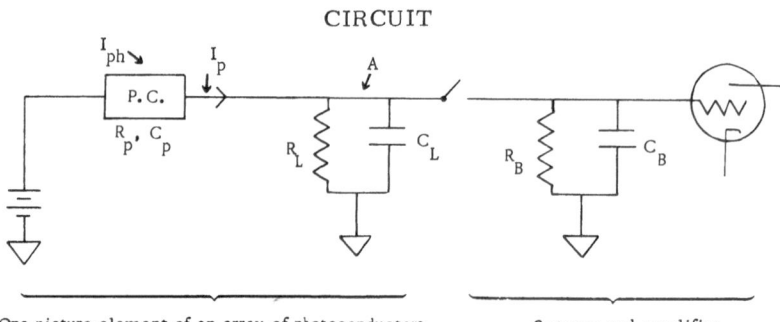

One picture element of an array of photoconductors   Scanner and amplifier

I_{ph} = photon sec strike photoconductor

τ_p = life time of photoelectron

$\tau_0 (\tau > \tau_p)$ = response time of photoconductor

STATEMENT OF PROBLEM

Figure 1 shows a photoconductor (P.C.) in series with a battery and a load resistance R_L and Capacitance C_L. The photoconductor has a dark resistance P_p and a capacitance C_p. R_L is the resultant of the actual

See also "An Analysis of Photoconductive Photon Counters" — A. Rose, 1969 Photoconductivity Conference (Stanford) to be published by Pergamon Press.

resistance in that part of the circuit and of R_P in parallel. Similarly, C_L is the sum of the actual capacitance and of C_P in parallel.

The point A may be connected permanently to an amplifier or may be sampled intermittently as in a scanned array. In a typical scanned array $R_L C_L$ has a time constant of about 0.1 seconds and $R_A C_A$ a time constant of about 10^{-7} seconds.

I_{ph} photons/sec strike the photoconductor. The free life time of the photoexcited carriers is τ_p and the response time of the photoconductor to changes in light intensity is τ_0. The response time is generally greater than the lifetime by the ratio of trapped to free carriers. The charge storage time $R_L C_L$ is matched to the response time so that

$$\tau_0 = R_L C_L$$

and $I_{ph}\tau_0$ electrons are excited during the response time.

The analysis will be carried out in terms of niose charges rather than noise currents or voltages. The noise charge approach allows a more graphic statement of the argument.

PHOTO-NOISE CHARGE AT A

Suppose the average number of electrons accumulated at A in the storage time is equal to the number of photons striking the photoconductor in time τ_0, that is, the photogain is unity. Then the r.m.s. fluctuation associated with $I_{ph} \tau_0$ is

$$(I_{ph} \ \tau_0)^{1/2}$$

and the photo noise charge is

$$e(I_{ph} \ \tau_0)^{1/2} .$$

Now let the photo gain be G. This means that the charge associated with each random event is Ge. Hence the photo noise is

$$Q_{PN} = Ge \ (I_{ph} \ \tau_0)^{1/2} .$$

The photo gain at fields high enough for space-change-limited-current flow is

$$G = \frac{\tau_0}{R_P C_P} = \frac{R_L C_L}{R_P C_P} .$$

THERMAL NOISE CHARGE

The noise charge at A due to Johnson noise in the resistor R_L can be computed as the product of noise current and storage time $R_L C_L$:

$$Q_{JN} = \left(\frac{4kT}{R_L} \Delta f \right)^{1/2} R_L C_L$$

$$= \left(\frac{4kT}{R_L} \frac{(R_L C_L)^2}{2\pi R_L C_L} \right)^{1/2}$$

$$= (2kTC_L)^{1/2} .$$

The condition for the dominance of the photo noise is

$$Q_{PN} \geq Q_{JN}$$

or
$$Ge(I_{ph}\,\tau_o)^{1/2} \geq (2kTC_L)^{1/2}$$

$$I_{ph}\,\tau_o \geq \frac{2kTC_L}{G^2 e^2}$$

$$\geq \frac{2kTC_L}{e^2}\,\left(\frac{R_p\,C_p}{R_L\,C_L}\right)^2.$$

For circuit reasons the smallest value of R_p/R_L is unity. Then, the threshold number of photons per picture element per observation time so that the photo noise $\geq$ Johnson noise is

$$N \equiv I_{ph}\,\tau_o \geq \frac{2kTC_p}{e^2}\,\frac{C_p}{C_L}$$

$$\geq 10^5\,C_p\,\frac{C_p}{C_L}\ \text{for } C_p \text{ in cms.}$$
$$\text{and } T = 300°\,K$$

$$\geq 10^5\,C_p/G\,.$$

For detection of single photons (N = 1)

$$1 \geq 10^5\,C_p/G$$

or
$$C_p \leq 10^{-5}G$$

$$\text{For } G = 1,\ C_p \leq 10^{-5}\ \text{cms.} \approx 10^3\ \overset{\circ}{A}.$$

SAMPLE CALCULATIONS

We begin with the following set of relations just derived:

$$C \leq 10^{-5}\,G\ \text{cms.}$$

$$\leq 10^{-17}\,G\ \text{farads}$$

$$\tau_o = R_L\,C_L$$

$$R_p = R_L = \frac{\tau_o}{C_L}$$

$$G = \frac{C_L}{C_p}\,.$$

We pick C_L as the independent variable since it must be $\geq C_B$ in order to avoid additional noise from the scanning process. τ_o is generally fixed at 10^{-1} sec in order to reproduce motion

$$C_p \leq 10^{-17}\,G = 10^{-17}\,\frac{C_L}{C_p}\,.$$

Hence: $$C_p \leq 3 \times 10^{-9}\ C_L^{\ 1/2}\ \text{farads}$$

and $$G = \frac{C_L}{C_p} = 3 \times 10^3\ C_L^{\ 1/2}\ .$$

For a cube of photoconductor of side d:

$$d = 10^{12}\ C_p = 3 \times 10^3\ C_L^{\ 1/2}\ \text{cm}$$

$$P = R_p d = \frac{3 \times 10^3\ \tau_o}{C_L^{\ 1/2}}\ \text{ohm cm.}$$

From the relation

$$G = \frac{\tau_o}{T} = \frac{\tau_o}{(d/\varepsilon\mu)}$$

we find

$$\varepsilon = \frac{G\ d}{\tau_o \mu} = \frac{10^{12}\ C_L}{\tau_o \mu}\ \text{volts/cm}$$

and

$$V = \varepsilon d = \frac{3 \times 10^{15}\ C_L^{\ 3/2}}{\tau_o\ \mu}\ \text{volts}\ .$$

153

VI. Ohm's Law and Hot Electrons

I. INTRODUCTION

Ohm's law is one of the first concepts that a student of physics encounters. It is presented as a kind of pin ball machine. The drifting electron repeatedly collides with the atoms of the solid after traversing a fixed mean free path. The mean free path is converted into a mean free time via the thermal velocity of the electron. From the mean free time and Newton's laws of motion the mobility

$$\mu = \frac{\tau_c e}{m} \tag{1}$$

is derived.

After this brief encounter with Ohm's law in his freshman year, the student is not likely to explore its mysteries again until some time in his graduate studies and only then if he undertakes a course in transport theory. At this time he learns that the electrons don't really collide with the atoms of the solid but rather that an electron in some energy state, characterized by the Schrödinger wave function Ψ_1, is perturbed by a phonon (also a wave motion) to make a transition to another energy state characterized by Ψ_2. The mean free time of the electron is obtained only after summing over all of the possible perturbations due to the spectrum of phonons. Moreover, the mean free time is no longer a constant of the material but depends on the energy of the electron. And, finally, the energy of the electron is a function of the applied field so that Ohm's law is only an approximation. Under certain conditions, major departures from Ohm's law can lead to both scientific as well as technological novelty.

Between the pin ball machine at one end and the formal sophistication of perturbation theory at the other, there is a time gap of some six years and a conceptual gap that is likely to be bridged only by a select group of solid state theorists. Certainly, there is a large and growing body of engineers and applied physicists who are intimately concerned with departures from Ohm's law in such phenomena as solid state microwave oscillators, solid state photo multipliers, dielectric breakdown and electroluminescence and whose exposure to Ohm's law falls considerably short of a graduate course in transport theory.

This discussion attempts to outline an intermediate approach to Ohm's law which might well be part of an undergraduate course of study. The approach is couched in real space (as opposed to phase space) and largely

Taken in part from: A. Rose, Classical aspects of Ohm's law. Helv. Phys. Acta, **41**, 1119 (1968).

in classical concepts. Quantum aspects of the problem are brought in at the end as constraints on the classical solution. At the expense of certain computational approximations, I believe that the approach yields a clear picture of the physics of electron phonon interactions.

II. OUTLINE OF ARGUMENT

The momentum of an electron interacting with the phonons of a solid is randomized in a time τ_c, called the mean collision time. While the electron absorbs and emits a spectrum of phonons, its collision time is dominated by the shortest wavelength phonons with which it can interact since the density of phonons increases as λ^{-3} and since the coupling to the shorter wavelength phonons is, in general, stronger.

The electron absorbs these phonons at a rate proportional to their density n_{ph} and emits them at a rate proportional to $n_{ph} + 1$. Also, each emission or absorption is sufficient to randomize the momentum of the electron. The term unity in the emission factor is called the spontaneous emission in contrast to that induced by the thermally generated density n_{ph}. The spontaneous emission is the rate at which an electron of finite energy loses energy via phonon emission in a zero temperature ($n_{ph} = 0$) lattice. The time to emit one phonon is then related to the average rate of loss of energy by

$$\frac{\hbar\omega}{\tau_e} = \frac{dE}{dt} \tag{2}$$

and the collision and emission times are related by

$$\tau_e = (2\,n_{ph} + 1)\,\tau_c$$

$$\doteq \frac{2kT}{\hbar\omega}\,\tau_c \text{ for } kT > \hbar\omega \tag{3}$$

At this point it is clear that the mobility of an electron can be computed directly by computing τ_c as is done in perturbation theory or by computing τ_e from Eq. (2) and converting to τ_c via Eq. (3), namely,

$$\tau_c = \frac{(\hbar\omega)^2}{2kT} \bigg/ \frac{dE}{dt} \tag{4}$$

The second approach to mobility depends essentially on being able to compute the average rate of energy loss dE/dt by an electron in a zero temperature lattice. A semiclassical method for carrying out this computation is outlined in the next section.

A knowledge of dE/dt also allows one to compute the mean energy (sometimes loosely called the temperature) of electrons that are heated above the lattice temperature by the applied field. The mean energy follows from the energy balance

155

$$\frac{dE}{dt} = \mathcal{E} ev_d = \mathcal{E}^2 e^2 \frac{\tau_c}{m} = \frac{(\mathcal{E} e\hbar\omega)^2}{2kTm} \quad \Big/ \quad \frac{dE}{dt}$$

or
$$\frac{dE}{dt} = \frac{\mathcal{E} e\hbar\omega}{m\,v_t} \tag{5}$$

Here dE/dt depends on the mean energy of the heated electron, $\mathcal{E}$ is the applied field and v_t the thermal velocity of an electron at the lattice temperature.

III. SPONTANEOUS EMISSION

Consider an electron immersed in a solid. Its coulomb field polarizes the solid medium so that the energy of the electron is lowered. For example, the energy per unit volume in the coulomb field in the solid is $\frac{1}{8\pi K}(e/r^2)^2$ while the same energy in a vacuum was $\frac{1}{8\pi}(e/r^2)^2$. The difference $\frac{K-1}{K}\frac{1}{8\pi}(e/r^2)^2$ is a measure of the work required to remove the electron from the solid to vaccum. It is also a measure of the energy-well the electron digs for itself by polarization of the solid.

If the electron moves sufficiently slowly, it carries this polarization along with it without loss of energy. At higher electron velocities response of the medium is not fast enough to keep pace with the electron and hence the polarization is left behind as a trail or wake of polarization energy [1]. This is, in essence, the classical physics of spontaneous emission. The magnitude of the rate of energy loss will be computed classically and certain quantum constraints will be imposed upon the solution to reconcile it with the laws of quantum physics.

Figure 1 shows the classical model on which the energy-loss computation is based. A particle moves past an array of elements with a velocity v. Each element has a dimension d and is attached to a spring such that it has a characteristic frequency ω. There is a force of attraction between the particle and each element such that when the particle is stationary an energy E_w^o is stored in the element opposite the particle. In this model, the particle represents the electron, the array of elements represent the medium, and the energy E_w^o corresponds to the energy-well formed by the interaction of electron and medium.

We can by inspection of Figure 1 approximate the maximum rate of loss of energy by the particle to the array of elements. Let the particle remain opposite an element for a time ω^{-1}, that is, for a time just sufficient to allow the element to respond and to store an energy E_w^o in its spring. Now move the particle abruptly to the next element. This will leave behind the energy E_w^o. The particle remains on the second element for a time ω^{-1} sufficient to allow it to respond and is then abruptly moved to the third element and so on. The rate of loss of energy is then

$$\frac{dE}{dt} = E_w^o\,\omega \tag{6}$$

and the average velocity of the particle is $v = d\omega$. It is clear that, within a factor of about 2, the same rate of loss of energy will be incurred if the step wise motion is replaced by the average velocity $d\omega$.

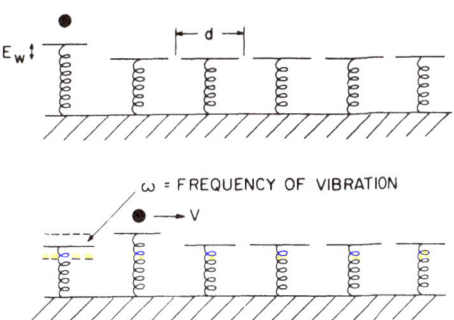

ω = FREQUENCY OF VIBRATION

FIGURE 1. Model for rate of loss of energy by electrons in a solid.

We wish now to compute the rate of loss of energy for velocities greater than $d\omega$. Under these conditions the particle passes each element in a time short compared with its response time and imparts a momentum to it proportional to its transit time. The energy imparted will be proportional to the square of the momentum or to the square of its transit time. Thus, the energy-well will be proportional to the same factor:

$$E_w = E_w^o \, (T\omega)^2$$
$$= E_w^o \left(\frac{d\omega}{v}\right)^2 \qquad (7)$$

The factor ω is a normalizing factor such that when the transit time $T \rightarrow \omega^{-1}$, the energy well $E_w \rightarrow E_w^o$ consistent with Eq. (6). To compute the time rate of loss of energy, we need to multiply Eq. (7) by the number of energy-wells traced out per second; namely,

$$\frac{dE}{dt} = E_w \frac{v}{d}$$
$$= E_w^o d \frac{\omega^2}{v} \qquad (8)$$

It remains to evalute E_w^o for an electron in a solid. To do so we recognize that the maximum energy the electron can impart to the medium energy is its coulomb energy $\frac{e^2}{d}$. Moreover, since we are concerned here with the interaction between the electron and the lattice ions or atoms, the available coulomb energy is reduced or screened by the electronic part of the dielectric constant to a value $e^2/(Kd)$ where K is the electronic part of the dielectric constant. The electron, for velocities not exceeding about 10 volts, carries this electronic polarization with it so that the lattice sees only the reduced coulomb field or energy. Since $e^2/(Kd)$ is the maximum energy the electron can impart to the lattice, we multiply it by a factor β, where $0 \leq \beta \leq 1$, to obtain E_w^o. By definition, β is the fraction of available coulomb energy of the electron that can be imparted to the lattice. With this substitution for E_w^o, Eq. (7) becomes

$$\frac{dE}{dt} = \beta \, \frac{e^2 \, \omega^2}{Kv} \qquad (9)$$

Eq. (9) is essentially the classical result for the rate of energy loss or the spontaneous emission by an electron of velocity v to phonons of frequency ω. The factor β is a coupling constant that is still to be evaluated and that is a function of the type of phonon as well as the frequency of the phonon. Eq. (9) was evaluated for the dimension d or rather for a spherical shell of dimensions $d \pm \dfrac{d}{2}$ surrounding the electron. A more complete analysis

[2] sums up the contributions from other shells so that Eq. (9) is to be multiplied by a numerical factor in the neighborhood of unity.

The major quantum constraint on Eq. (9) is that imposed by the quantum rule of K-conservation. The shortest wavelength phonon an electron of velocity v can emit is given by

$$\lambdabar \equiv \frac{\lambda}{2\pi} \approx \frac{\hbar}{2mv} . \tag{10}$$

This constraint can also be looked upon semiclassically as saying that the uncertainty radius of the electron should not exceed the wavelength of the phonon it emits. In the case of acoustic phonons Eq. (10) can be rewritten as

$$\omega = \frac{2\,m\,v\,v_s}{\hbar} , \tag{11}$$

where v_s is the velocity of sound. By Eq. (11) the appropriate ω in Eq. (9) is defined for each velocity v of the electron.

It remains to clarify the meaning of the coupling constant β. This has been done in detail in [2] and [3]. We quote here the results. On the one hand β was defined as the fraction of available coulomb energy of the electron that can be transferred to the medium or, more particularly, to the mode of vibration of the medium through which the electron loses energy. On the other hand, it can be shown [2, 3] that this definition of β is equivalent to the ratio of electrical to total energy of the macroscopic sound wave corresponding to the type of phonon or electron-phonon coupling under consideration. For example, in the case of acoustic phonons coupled to electrons via their piezoelectric fields, β is essentially the well known quantity, the square of the electromechanical coupling coefficient, long used to analyse crystal oscillators. In general, β has an easily visualizable physical meaning for the several types of phonons and electron-phonon couplings. Its values are listed in Table I. This form of coupling constant has the additional virtues that it is valid also for electron-electron interactions and for identifying the several types of acoustoelectric interactions. In brief, it unifies an extensive body of otherwise disconnected literature.

TABLE I.

Phonons	Coupling	β	v-dependence of dE/dt
Polar Optical	Polarization field	$\dfrac{\epsilon_o - \epsilon_\infty}{\epsilon_o}$	v^{-1}
Non-Polar Optical	Deformation Potential (D)	$\dfrac{KD^2}{4\pi\rho e^2\omega^2 \lambda^2}$	v
Acoustic	Piezoelectric field	$\dfrac{\epsilon_p^2}{KC}$	v
Acoustic	Deformation Potential (B)	$\dfrac{KB^2\omega^2}{4\pi\rho e^2 v_s^4}$	v^3

Definitions:

ϵ_o low frequency dielectric constant

ϵ_∞ high frequency (optical) dielectric constant

K dielectric constant at frequencies $\gtrsim \omega$

ϵ_p piezoelectric constant

C elastic constant

ρ density (grams cm^3)

v_s velocity of sound

IV. SAMPLE CALCULATIONS

The utility of Table I together with Eqs. (2-5), (10) and (11) is illustrated by the following sample calculations.

Eqs. (2-5) can be written as

$$\frac{dE}{dt} = \frac{\hbar\omega}{\tau_e} = \frac{(\hbar\omega)^2}{2kT\,\tau_c} = \frac{\mathcal{E}e\hbar\omega}{mv_t}$$

and Eqs. (10) and (11) as

$$\hbar\omega = 2\,m\,v\,v_s = 2\,m\,v\,\lambda\,\omega .$$

In thermal equilibrium $v = v_t$ and the temperature dependence of mobility for interaction with acoustic phonons becomes

159

$$\mu = \frac{\tau_c e}{m} \infty \frac{\omega^2}{v_t^2} \Big/ \frac{dE}{dt}$$

$$\infty \begin{cases} v_t^{-3} \infty T^{-3/2} & \text{deformation potential coupling} \\ v_t^{-1} \infty T^{-1/2} & \text{piezoelectric coupling} \end{cases}$$

The dependence on electric field of the mean energy E of hot electrons is obtained by

$$\frac{1}{\omega} \frac{dE}{dt} \infty \mathcal{E}$$

or $\begin{cases} v^2 \infty E \infty \mathcal{E} & \text{deformation potential coupling to acoustic phonons} \\ v \infty E^{1/2} \infty \mathcal{E} & \text{deformation potential coupling to optical phonons} \end{cases}$

The threshold field for generating hot electrons when the electrons interact with acoustic phonons is obtained from

$$\frac{(\hbar\omega)^2}{2kT \tau_c} = \frac{\mathcal{E} e\hbar\omega}{mv_t}$$

or

$$\frac{\mathcal{E} e\tau_c}{m} = v_{drift} = \frac{\hbar\omega}{mv_t} = 2 v_s$$

Threshold field for hot electrons and for runaway to dielectric breakdown calculated for interaction with polar optical phonons.

$$\mathcal{E} e\, v_{drift} = \beta \frac{e^2 \omega^2}{\varepsilon_\infty v_{random}}$$

At breakdown: $v_{drift} = v_{random} \equiv v$

Hence,

$$\mathcal{E} = \frac{\beta}{\varepsilon_\infty} \frac{e\, \omega^2}{v^2}$$

Also, at breakdown $\frac{1}{2} mv^2 = \hbar\omega$

Hence, $\mathcal{E} = \frac{\beta}{2\varepsilon_\infty} \frac{em\omega}{\hbar}$ esu/cm

$$= 150 \frac{\beta}{\varepsilon_\infty} \frac{em\omega}{\hbar} \text{ volts/cm}$$

$$= 150 \frac{\varepsilon_0 - \varepsilon_\infty}{\varepsilon_0\, \varepsilon_\infty} \frac{em\omega}{\hbar} \text{ volts/cm}$$

REFERENCES

1. H. FROHLICH and E. PELZER, ERA Report #L/T 184 (1948)
2. A. ROSE, RCA Rev., 27, 600 (1966)
3. A. ROSE, RCA Rev., 27, 98 (1966)

5/Principles of Solidification

B. CHALMERS

I. INTRODUCTION

Solidification, in the sense used in this context, is the process by which a liquid is transformed into a crystalline solid. In crystal growth the solid that forms first is solvent rich as distinct from crystallisation, in which the crystals that are formed are solute rich. It is not always possible to make a clear distinction. Solidification is important as the process employed in the widely used process of casting, in all its forms from large ingots of steel to small crystals of silicon. While in principle it would seem simple to convert a homogeneous liquid into an equally homogeneous perfect crystal, this is extremely difficult, if not impossible to achieve in practice. Thorough understanding requires that the process be studied at various levels, which can be conveniently described as the angstrom level, the micron level and the centimetre level.

At the angstrom level we must consider the atomic configuration of the interface between the solid and the liquid, and the atomic movements that take place there, both while the crystal is growing and while it is not, and we must examine the very first step in the growth of a crystal, the formation of a viable nucleus.

At the micron level, it is necessary to examine the redistribution of solute that is required when the crystal that is formed has a different composition from the liquid. This can cause unwanted segregation or highly desirable purification in a solid consisting of more than one component and we must consider the conditions under which the interface between solid and liquid becomes geometrically unstable and gives rise to cellular or dendritic structures; the origin of various kinds of imperfections in the resulting crystals is also of interest.

The study of solidification at the centimetre scale of sizes is of interest mainly in connection with applications rather than principles; it includes the study of heat flow and fluid flow in a solidifying liquid, and the control of the location of the voids that occur when the solid occupies a smaller volume than the liquid.

II. THE INTERFACE

The unit process is one atom going from liquid to solid or solid to liquid. It is the movement of an atom over an energy saddle-point from a position of higher to one of lower energy (or vice versa). This means that (from liquid to solid) the atom is attracted to a more strongly bound (lower energy) position; it accelerates and when it reaches its new position, it may rebound elastically or the collision may be inelastic, in which case the excess kinetic energy is shared with surrounding atoms. This is the latent heat of the transformation. It diffuses outwards and locally decreases the driving force (or accelerates the reverse process).

Each process (liquid to solid and solid to liquid) is thermally activated in the sense that higher than average thermal energy is required to make the transition. The net rate is the difference between the two individual rates, each governed by a law of the form $A \exp(-Q/RT)$; where A is a geometrical constant, including a term for the accommodation coefficient (the probability of an inelastic collision); Q is the activation energy for the process in question. At equilibrium, $A_1 \exp(-Q_1/RT) = A_2 \exp(-Q_2/RT)$ Q_1-Q_2 is the latent heat per atom.

The value of the accommodation coefficient for freezing depends on the nature of the interface, which may be "rough", "smooth" or intermediate. If rough, all sites are equally available, and the accommodation coefficient is high. If smooth, in the atomic sense, the only available sites are on steps, or even at step kinks. In such cases the value of A_1 is much lower. This means that the "freezing" process is relatively slow and the melting point is relatively low, and the entropy of melting is high. In some cases the nature of the interface changes from smooth, at low driving force, to rough for high driving force. In general, metals have low entropy of melting and a rough interface, while ionic and covalent solids have smooth interfaces. When the interface is rough, the morphology is controlled by heat flow, when smooth, by layer growth. However, even in the case of the rough interface, the net kinetics apparently have some anisotropy (dendritic growth is crystallographic). Kinetics are effectively linear near the melting point.

The net rate of freezing or melting is proportional to the local departure from equilibrium. The actual rate is controlled by heat flow, as the latent heat always goes towards reducing the driving force.

It can easily be shown that the kinetic and thermodynamic descriptions of the melting point are equivalent.

III. NUCLEATION

This is the process by which a new crystal is formed; the difficulty in forming a new crystal is that the free energy of a crystal includes that due to its surface, as well as the volume free energy. When a crystal is very small, the surface term is relatively large; a small crystal is, therefore, in equilibrium with the melt (i.e. equal free energy) at a temperature below the melting point. For any given radius of curvature, there is a temperature of equilibrium with the melt. This equilibrium is unstable. The radius of unstable equilibrium is the critical radius r*. Relationship between r* and ΔT is given in Figure 1. This determines how large a crystal must be (at any given temperature), to have equal probability of growing or melting.

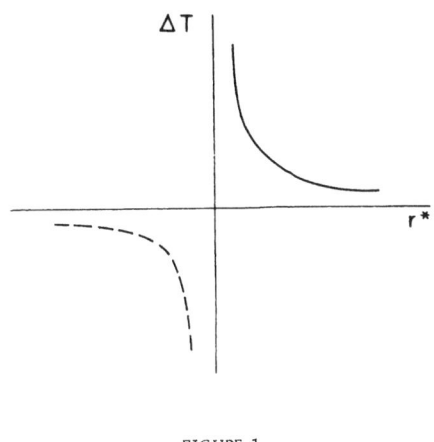

FIGURE 1

The second component of the problem is how large an embryo is likely to form spontaneously in the melt. An embryo is a cluster of atoms whose arrangement is that of the crystal; it exists for a very short time, because of the very rapid atomic processes at its surface, unless it reaches critical size and grows. Detailed study of the kinetics of growth and melting show that, for any temperature, there is a maximum size of any embryo that is likely to occur. This size can be described in terms of its radius, r_E, although the real criterion is the number of atoms in it. Nucleation occurs when R_E equals $R*$ at the temperature ΔT_N.

For metals, this can be expressed non-dimensionally (as shown in Figures 2 and 3):

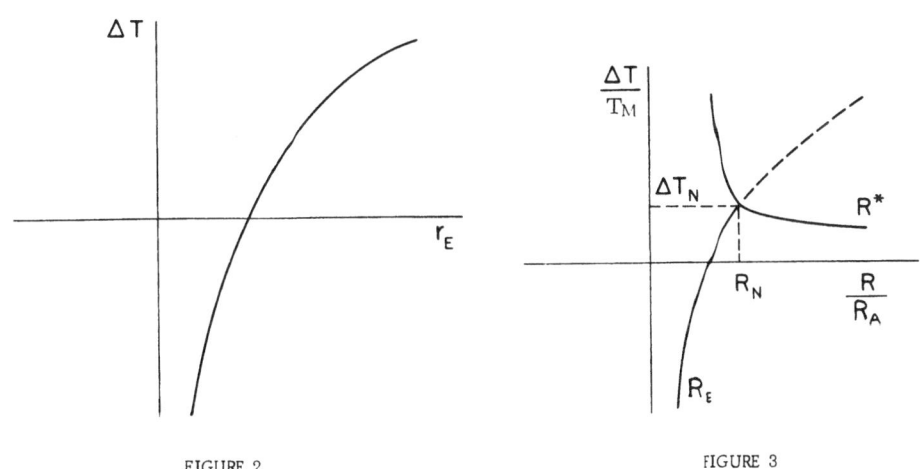

FIGURE 2   FIGURE 3

$$\text{and} \quad \frac{\Delta T_N}{T_M} \approx 0.2; \quad 3 < \frac{R*}{R_A} < 4$$

where T_M is the melting point, R_A is the atomic radius, $R*$ critical radius of nucleation and ΔT_N the nucleation temperature gradient. Homogeneous nucleation seldom occurs because another process, heterogeneous nucleation, usually intervenes before $\Delta T_N / T_M$ reaches the required value. Heterogeneous nucleation is the process in which an embryo of critical radius forms on a substrate, which allows the required critical radius to be reached with fewer atoms than would be required for a complete "spherical" nucleus. This occurs in two ways:

1. The nucleus approximates to a spherical cap (Figure 4); this requires that the interfacial free energy between crystal-substrate be less than the interfacial free energy liquid substrate (Figure 4). This changes the relationship between the volume and the radius of the embryo. The diagram (Figure 5) shows that the "cap" embryo becomes critical at less ΔT than the "spherical" embryo.

2. The size of nucleus is such that, for very effective heterogeneous nucleation, the "cap" must be a monolayer of the crystalline phase on the substrate. It is no longer realistic to use the term "radius". The criterion becomes one of adsorption of a layer of atoms of the liquid phase on the substrate in a pattern on which the crystal can grow. This requires that the adsorption sites must match, within fairly close limits, the sites of an atomic plane of the crystal. This is not sufficient; the surface of the substrate must be flat, atomically, over an area of sufficient size that a "patch" of critical size is likely to form on it.

A liquid can exist indefinitely at any temperature above that at which nucleation takes place; however, vibration of sufficient intensity to produce cavitation can cause nucleation even when the liquid is in the metastable range. The effect of cavitation is believed to depend on the change of melting point with pressure, a maximum of which is reached during collapse.

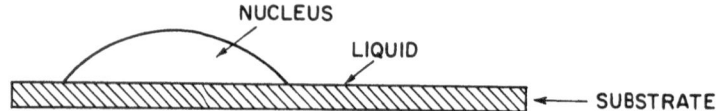

FIGURE 4

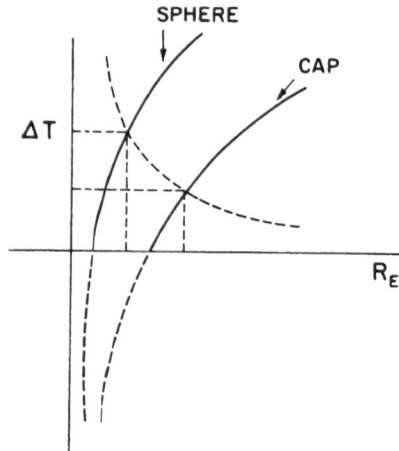

FIGURE 5

IV. MORPHOLOGY OF THE GROWING CRYSTAL

The morphology of the growing crystal (i.e. the micro geometry of the solid-liquid interface) depends on two factors: 1) The direction and gradient of the driving force, which in the case of crystal growth from the melt is the local undercooling of the melt. The undercooling may increase (undercooling gradient positive) or it may decrease (undercooling gradient negative) with distance from the interface into the melt, and 2) The kinetics of the growth process: in particular, whether the interface is "rough", in which case growth takes place in the normal direction, or "smooth", when growth is predominantly by the lateral propagation of steps. There are, in general, four cases:

1) Rough interface, undercooling gradient negative. This is the normal case when heat is extracted through the solid, this heat being the sum of the latent heat of fusion and that which reaches the interface by conduction down the temperature gradient. If the liquid is pure, the shape of the interface matches the shape of the isothermal surfaces, the only departure being due to the anisotropy of the interface kinetics which, for a rough interface, is relatively small. The isothermal surface is controlled by the geometry of the system and can be distorted by anisotropy of thermal conduction in the solid.

2) Smooth interface, undercooling gradient negative. In this case, the interface tends to be flat, but if the appropriate crystallographic face is inclined to the isothermal surface, a compromise is reached between the two factors. The surface must remain between two isothermals, and this may cause more faces to appear.

3) Rough interface, positive gradient. This case can arise when crystal growth takes place in an undercooled melt. It is the most interesting case, because a flat surface is unstable in these circumstances. Growth is, in general, heat flow controlled, and as a consequence, any convex perturbation tends to increase. This gives rise to dendritic growth. However, at very small radii the effect of convexity on heat flow is counteracted by the decrease of equilibrium temperature that corresponds to high curvature: if the crystal is spherical, for example, it is stable in shape until its radius reaches about seven times its critical radius; then perturbations can have large enough radii to survive. It is not known whether growth is a steady state process, with a stable shape, or whether it is an inherently fluctuating process, the branches forming as an inherent part of the process. The characteristic direction of growth of dendrites must be caused by anisotropy of kinetics, although the interface is rough; it is possible that certain faces (the most closely packed) are sufficiently smooth to control the growth direction, which is always the axis of a pyramid bounded by the most closely packed planes that are available.

4) Smooth interface, positive gradient. In this case the crystal is bounded by low index crystallographic planes, but steep gradients of undercooling can produce shapes which at insufficient resolution look like dendrites; an interesting case is that in which the initiation of new layers on the surface is facilitated by a crystallographic feature, for

example, a twin. In this case the morphology is controlled by the generation of new steps, while the rate of growth depends on the rate at which heat is conducted away into the liquid and into the crystal.

V. REDISTRIBUTION OF SOLUTE

For almost all alloy compositions, the crystalline phase has a different composition from the liquid with which it is in equilibrium. This is a consequence of solute atoms, in the crystal, having different, usually lower, activation energy for escape than the solvent atoms. In such cases, the solid-liquid equilibrium temperature of the alloy is lower than that of the pure solvent, and the crystal contains less solute than the liquid. The ratio of the equilibrium compositions is called the distribution coefficient, or partition coefficient, K_0. It is usually reasonable to assume that liquidus and solidus lines are straight, in which case K_0 is independent of composition. When conditions are such that the interface is planar and the liquid at rest, the redistribution of solute during solidification is easily computed. The solute "rejected" at the interface diffuses normally to the interface into the liquid. The final distribution is as shown in the diagram of Figure 6.

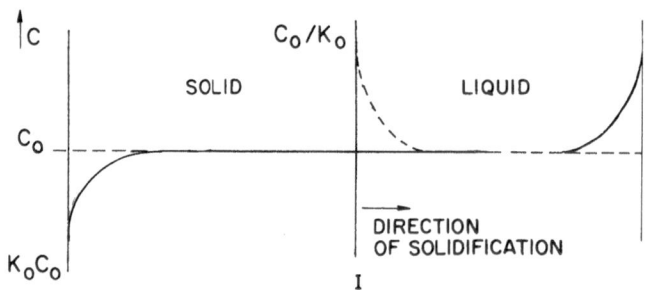

FIGURE 6

The instantaneous distribution of solute when the interface is at I is also shown. If the liquid is stirred, so that its composition is, at any moment, uniform, the composition profile becomes as shown in Figure 7.

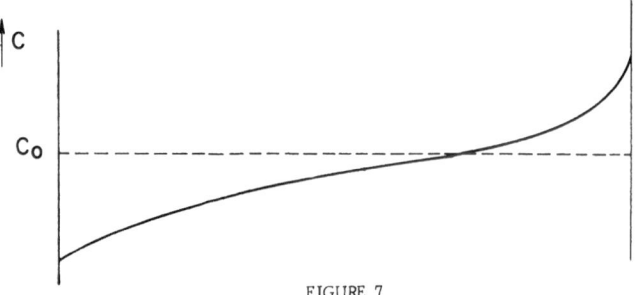

FIGURE 7

This increases the transfer of solute. Zone refining uses this principle. Each molten zone deposits solid of higher purity than the liquid in it, and so transfers more solute from left to right in the diagrams in Figures 8 and 9.

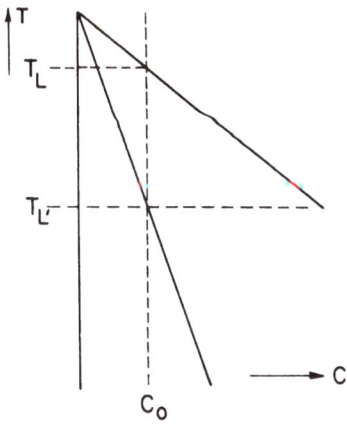

FIGURE 8

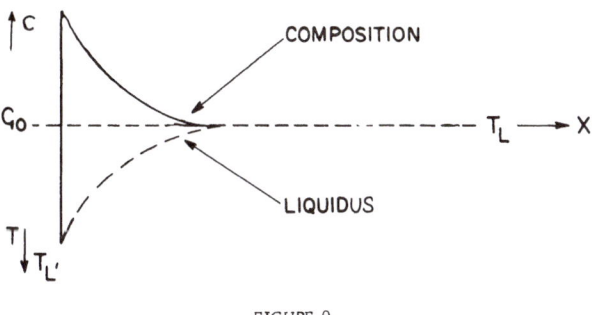

FIGURE 9

With a planar interface, the composition ahead of the interface is such that the local liquidus temperature rises with increasing distance from the interface. The temperature of the interface is below the liquidus temperature corresponding to C_0. If the temperature gradient in the liquid is not too large some or all of the liquid is undercooled (constitutional undercooling) and the driving force for solidification increases with distance from the interface (i.e. positive gradient of undercooling) as shown in Figure 10. This causes the planar interface to be unstable. This can have any of the three consequences:

1) If the temperature gradient in the liquid is such that the layer of the supercooled liquid is thin, a cellular segregation pattern develops (Figures 11 and 12).

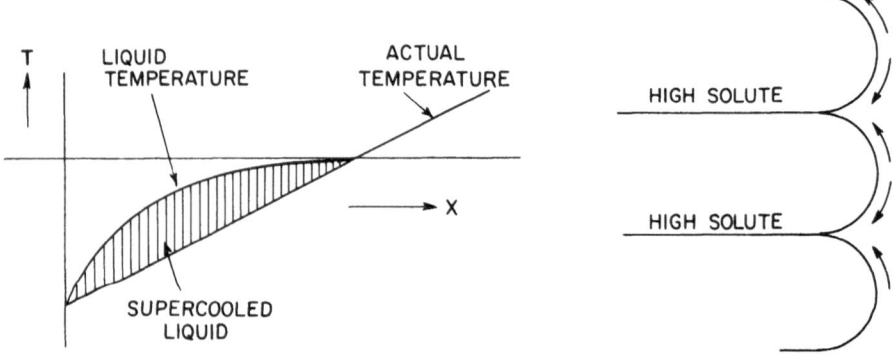

FIGURE 10   FIGURE 11

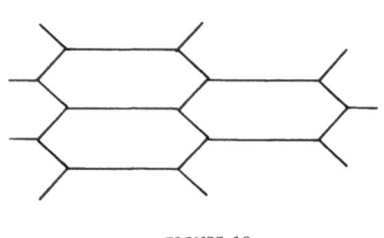

FIGURE 12

This minimizes undercooling. The linear model is no longer valid, since there is lateral diffusion of solute. This structure can be suppressed by slow growth, high purity or a steep temperature gradient in the liquid.

2) If the undercooling extends a long way into the liquid, the instability can develop into a dendritic structure. It is important to note that a dendritic structure can be formed in a solution, even if there is heat flow from the liquid to the solid. The dendritic structure is somewhat different from free dendrites that grow in a thermally supercooled melt. The alloy dendrites cannot extend beyond a limiting isotherm, whose rate of movement determines the rate of growth of the dendrites. Dendritic growth causes very substantial transverse diffusion, with strong segregation effects. The temperature of the interface (the limiting isotherm) is only slightly below the bulk liquidus temperature because outward diffusion from dendrite tips is much more effective than purely forward diffusion ahead of a planar interface.

3) If the constitutional supercooling is sufficient, it may cause the nucleation of new crystals ahead of the existing interface. This happens most frequently when a good nucleant is present. If the nucleation temperature is above the steady state growth temperature of the dendritic structure, nucleation will occur frequently and a small crystal size will result.

VI. ORIGIN AND CONTROL OF IMPERFECTIONS

The ideal crystal is one in which each atom (or ion) is in its place on a lattice that extends throughout the entire crystal. This implies complete purity (for an element) or precise stoichiometry (for a compound) or uniform composition for a solid solution; it also implies the absence of point defects (vacancies and interstitials), line defects (dislocations), surface defects (stacking faults), twin boundaries, boundaries between regions of different composition, antiphase boundaries in ordered structure and domain boundaries in ferromagnetic substances; and it could be taken to require the absence of thermal motion.

Imperfections are of two types: those that are in thermal equilibrium, and those that increase the free energy of the crystal. In the former category are thermal motion and point defects; all the other types of imperfection are present as a result of the constraints imposed by the conditions of growth or subsequent treatment and not by thermodynamic necessity. Each type of imperfection will be considered separately.

1) Point defects. Decreased temperature necessarily leads to supersaturation of vacancies; these can condense in various ways; voids in the interior or pits at the surface; discs which may collapse into dislocation loops or partial dislocation loops surrounding a disc of stacking fault, or stacking fault tetrahedra, or they may cause dislocations with edge components to climb. In general, the energetically most favourable, but slowest, process is "drainage" to the surface.

2) Dislocations. It is possible to grow dislocation free crystals; this seems to depend on maintaining all stresses, whether caused thermally, mechanically or chemically, at a very low level. This suggests that dislocations may always form. If they are caused to multiply by the action of a stress, they remain in the crystal, but if the stresses are so low that this does not occur, then they are able to move out of the crystal by climb or glide. This leaves the question of where they came from; a possibility is that there always are dislocation loops caused by vacancy condensation, which may be undetectable with the techniques used. A stress could cause loops to grow, multiply, become visible and be trapped in the crystal by the formation of a metastable network. Very slow cooling would minimize the formation of dislocation loops, if these are caused by vacancies. It seems clear that dislocations cannot be nucleated by strain, even at the melting point. The suggestion that vacancies trapped by the growth process may contribute is untenable.

3) Grain boundaries. In practice, the problem is the presence of sub-boundaries, that separate regions whose orientations differ by small angles (up to a degree or so). These often form by the aggregation of dislocations until a stable angle (about one degree) is reached. The dislocations are edge dislocations whose axis is perpendicular (or nearly so) to the interface and whose Burgers Vectors are parallel to it, and of which both ends are on the interface. Their origin is still obscure.

4) Stacking faults. There are boundaries between regions that have the same orientation but are out of alignment or registry. They can be formed by the motion of partial dislocations, and can occur as a feature of growth. In the latter case there are two possibilities: a) the collapse of a vacancy disc to form a partial dislocation loop surrounding a stacking

fault, b) an "error" in the positions taken by atoms as they attach to the surface of the crystal. There is no evidence that either of these processes actually occurs.

5) S e g r e g a t i o n .   The most drastic aspect of redistribution occurs at the end of the freezing process. The terminal transient can cause a very large increase in solute content, so that the obvious conclusions from a phase diagram can completely fail to predict the limiting composition and the temperature at which solidification finishes.

Fluid motion always tends to reduce the effects discussed above; but there is always a significant part of the diffusion layer within the hydro-dynamically stagnant boundary layer, so that constitutional supercooling cannot be suppressed completely even by forced motion of the liquid.

In extreme cases, segregation of a solute can, as noted previously, cause the nucleation of a second phase; this may be one component of a eutectic or peritectic or it may be a completely separate phase, either solid or gaseous; in the latter case pores or blow holes may be produced. This can usually be avoided by allowing the excess gas to diffuse away from the interface to a free surface. Since a gaseous solute need not be conserved, a "terminal transient" of very high concentration can be avoided.

6) P a r t i c l e s .   There is evidence that solid particles that are chemically inert with respect to the liquid can either be "pushed" by the advancing interface or "trapped". The speed of advance of the interface and its topography are determining factors, and the surface characteristics of the particle may also be significant.

6/ Techniques of Crystal Growth

R. A. LAUDISE

I. INTRODUCTION

Single crystals are the sine qua non of most solid state research and are an essential element in many solid state devices, and yet nowhere near as much orderly thought has been devoted to their preparation as might be expected. To the neophyte crystal grower there seems to be a myriad of unrelated growth techniques with few relationships among them, and the choice of a technique for a particular material appears to hinge more on whim and good luck than on any systematic rationale. The main object of this paper is to outline the various methods of crystal growth, to present their advantages and disadvantages, contrast and compare the various methods and to illustrate them with recent materials which have been grown, all with a view to aiding the grower in choice of method. This written report of the lectures is meant to serve only as a brief outline of the material presented.

A scheme which the author has found useful in categorizing growth methods is shown in Table I. In monocomponent methods the only component (component is used here in the same sense as in phase equilibria studies) present is that which will form the desired crystal. Traces of undesired impurities or deliberately added dopants are ignored as long as their presence doesn't result in crystal growth being severely limited by diffusion processes. In polycomponent methods an additional component or components have been added usually to either depress the melting point (growth from solution) or increase the volatility (growth by vapor phase reaction) of the material to be crystallized. Whenever a crystal is formed, an equilibrium must be shifted so as to favor the slow controlled formation of a solid crystalline product. The equilibrium may be either solid-solid, liquid-solid or gas-solid. Solid-solid polycomponent growth includes exsolution and solid state precipitation which have not been used for crystal formation, and so will not be discussed here. The main reason for the use of polycomponent growth is to permit crystallization at a lower temperature. Among the reasons why lower temperature growth is often desired are:

    1. To avoid high temperature polymorphs.

    2. To avoid decomposition and incongruent melting.

    3. To reduce vacancy concentration, strains, dislocations, and low angle grain boundaries.

    4. For experimental convenience.

TABLE I. Techniques of Crystal Growth

I. Monocomponent
   A. Solid-solid (recrystallization)
      1. Strain-annealing
      2. Devitrification
      3. Polymorphic phase change
      4. Sintering
   B. Liquid-solid
      1. Conservative
         (a) Directional solidification (Stockbarger-Bridgman)
         (b) Cooled seed (Kyropoulos)
         (c) Pulling (Czochralski)
      2. Nonconservative
         (a) Zoning (horizontal, vertical, float zone, growth on a pedestal)
         (b) Verneuil (flame fusion, plasma, arc image)
   C. Gas-solid
      1. Sublimation-condensation (vacuum evaporation, sputtering)
II. Polycomponent
   A. Solid-solid
      1. Precipitation from solid solution (exsolution)
   B. Liquid-solid
      1. Growth from solution (evaporation, slow cooling temperature differential, and solution transport-thermal gradient zone melting)
         (a) Aqueous solvents
         (b) Organic solvents
         (c) Molten salt solvents
         (d) Solvents under hydrothermal conditions
         (e) Other solvents (metals)
      2. Growth by reaction (media as above - temperature change, concentration change)
         (a) Chemical reaction
         (b) Electro-chemical reaction
   C. Gas-solid
      1. Growth by reaction (temperature change, concentration change)
         (a) Van Arkel
         (b) Epitaxial
         (c) Gas phase growth of inorganics
   D. Composite (vapor-liquid-solid)

The difficulties associated with polycomponent growth come about mainly because the added component(s) must diffuse away from the growing interface and the component forming the crystal must diffuse toward it. Concentration gradients associated with diffusion processes lead to constitutional supercooling, facet effects, cellular growth and dendritic growth. In addition, second components will have solid solubility in the grown crystal, and their presence may adversely affect its properties. Irregular convection in polycomponent systems can lead to temperature fluctuations which cause rate variations and lead to impurity banding. Polycomponent growth processes generally must proceed at slower rates to yield grown crystals of comparable perfection to monocomponent processes.

II. GROWTH BY SOLID-SOLID TRANSFORMATION

Solid-solid transformations are usually rather difficult to control as crystal growth methods, since the density of sites for nucleation in the transforming solid is so high it is difficult to cause nucleation and growth to occur at a single or even at a few sites. Consequently, the usual result of solid-solid growth is a polycrystalline product. Even with careful control, low angle grain boundaries, twins, and stacking faults are hard to avoid. The main advantages of solid-solid methods are that growth usually occurs at comparatively low temperatures and that the geometry of the grown crystal is fixed by the geometry of the specimen beforehand. Thus, single crystal wires and foils may be prepared.

Growth by polymorphic transformation has been used to prepare crystals of low temperature polymorphs when crystals of the high temperature polymorph are prepared by melt growth. The perfection of the resultant low temperature polymorph is high only when the two structures are quite similar. Transformation by moving a temperature gradient through the sample (with a geometry like that used in Stockbarger-Bridgman growth) would be particularly effective in controlling nucleation. The principal driving forces for recrystallization by solid-solid transformation are:

1. Energy differences due to grain size – large grains tend to consume small grains.

2. Energy differences due to orientation – grains bounded by low specific surface free energy faces tend to consume those with high specific surface free energy surfaces.

3. Strain energy introduced into the specimen by plastically deforming it.

Driving forces (1) and (2) are responsible for grain growth by sintering. Driving forces (1), (2) and (3) are responsible for recrystallization by strain-annealing. The most powerful solid-solid transformation technique is strain-annealing, and it has generally been most effective with metals because they are rather easy to plastically deform. The rupture strength of non-metals lies so close to their plastic deformation limit that very little strain energy can be introduced, so strain-annealing is not very effective for them. After the strain is introduced, the temperature is raised (annealing) to increase the rate of the solid state diffusion steps which are responsible for grain growth. Sintering must take place at elevated temperatures for the same reason. Although little used, temperature profiles and specimen shapes similar to those employed in Stockbarger-Bridgman growth should prove effective in localizing initial nucleation.

Some success in strain-annealing inorganics has been achieved by compressively loading specimens (so as to introduce strain energy) and annealing at the same time. This allows considerable strain energy to be introduced in spite of the fact that the rupture strength is close to the plastic deformation region. $ZnWO_4$ and Al_2O_3 crystals of several mm to nearly 1 cm have been grown by this method, and it may have considerable general utility.

III. GROWTH FROM THE PURE MELT

Growth by liquid-solid equilibrium is probably the most powerful and
generally useful growth technique for materials which melt without
decomposition. Melt techniques are usually divided into conservative and
nonconservative methods. In conservative methods, nothing is added to
or taken from the melt except by the freezing process. Stockbarger-
Bridgman and Czochralski growth are typical examples. In non-
conservative methods, new material can be added to the melt as growth
proceeds as in zone melting and Verneuil growth. The impurity
distribution which results in conservative and non-conservative methods
is quite different and has been discussed in the literature (Thurmond, 1959).
Another important distinction in methods depends upon whether a crucible
is required. Float zoning and Verneuil growth, the two main crucibleless
methods, are especially useful when an unreactive crucible for the molten
material cannot be obtained. However, since most crucibleless methods
depend upon surface tension supporting a molten material upon solid
material, there are constraints as to the geometry of crystals which can
be grown by such methods.

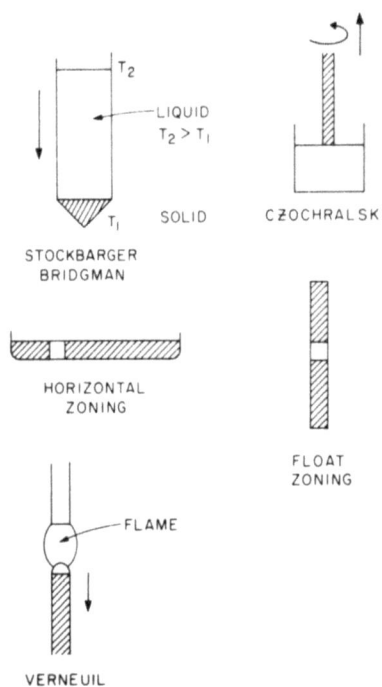

FIGURE 1. Melt growth methods

The dependence of distribution
constant for a given impurity upon
the nature and concentration of other
impurities present has often been
overlooked. Nassau's (1963) work on
$CaWO_4$:Nd and $CaWO_4$:Nd,Na is
instructive in this regard.

Czochralski growth is perhaps the
most generally useful of the pure melt
growth techniques. If the melt
material is compatible with the
crucible, few other constraints are
imposed on the type of materials
which can be grown. Growth can be
observed while in progress, high
perfection crystals can be obtained by
suitable technique modifications
(dislocation free Si and Ge are the
most noteworthy examples), and
little a priori knowledge of the system
is required. Indeed, crystal pulling
has proved a very powerful tool in the
search for nonlinear optical crystals,
where nucleation on a noble metal
wire in the absence of seeds has been
shown to give very useful preliminary
information about the amenability of
materials to growth by pulling. High
melting materials such as sapphire
and yttrium aluminum garnet are now routinely grown by Czochralski
growth from Ir crucibles.

The push toward higher melting point materials has focused considerable recent interest upon crucibleless techniques and high temperature heat sources. Hollow cathode heated float zoning and the replacement of arc image heaters with high intensity gas lamps, both show great promise of opening up materials with m. p. 's above 2500°.

IV. GROWTH FROM THE VAPOR PHASE

Growth from monocomponent vapor is generally by two methods:
1. Vacuum evaporation (sublimation in vacuo)
2. Sputtering

Few materials have high enough vapor pressures to make growth by sublimation feasible. However, where applicable (which so far has been mostly for metals) the technique is quite useful for preparing thin films. The usual product is a polycrystallic film, but provided the surface mobility of species which strike the seed is high enough, single crystals can be grown. To increase surface mobility, substrate temperature generally must be rather high if single crystals are desired. When the vapor pressure of a material is not high enough, the concentration of volatile species may be increased by placing the source in an electric field and "bombarding" it with appropriate ions. This technique is called sputtering and has been mainly applied to the preparation of polycrystalline films. However, if the seed temperature is made high enough, single crystal films can be prepared. Sputtering takes place in a low partial pressure of a (usually) inert gas such as Ar. The source is the cathode and the seed the anode. Ar^+ ions are produced which then strike the cathode and dislodge changed species which move to the seed. If, instead of an inert gas, a reactive gas is used (reactive sputtering) compounds can be grown. For instance, with a Zn source and O_2 as the gas, ZnO films with considerable single crystallinity can be prepared.

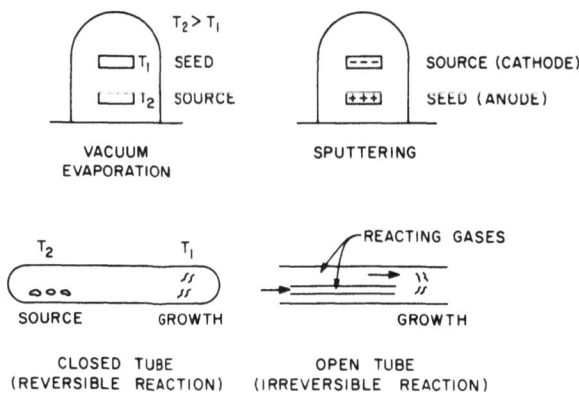

FIGURE 2. Vapor growth methods

If the material to be grown has too low a volatility, it can sometimes be complexed with an appropriate reactant so as to form a volatile species. Often at an appropriate temperature the reaction can be reversed causing the deposition of the desired material as a single crystal. This technique is termed growth by reversible reaction. If a reversible reaction cannot be found, then two volatile reactants which form the desired solid can be mixed close to the desired reaction site. This is growth by irreversible reaction. Reversible reactions are usually run in a closed tube with the complexing reagent returning to the source after deposition to pick up more material for deposition. Irreversible reactions are generally run in an open tube arranged with appropriate inlets for reacting gases and outlets for reaction products and unreacted gases. Reversible reactions are usually easier to control (usually this is done by controlling the temperature difference, ΔT, between source and seed) and result in better growth. For a reaction to be reversible and controllable

1. $\Delta G°$ for the reaction should be ≈ 0
2. $\Delta H°$ should be such that an appropriate ΔT shifts the equilibrium toward the formation of the desired crystalline product.

Thermodynamic data for many possible gas phase reactions are available so that polycomponent vapor growth is one of the few techniques where *a priori* calculations of feasibility are possible. If an irreversible reaction is used, it may be possible to grow good crystals on seeds if the seed temperature is made quite high so as to maximize surface mobility as has been done in the growth of sapphire.

V. GROWTH FROM SOLUTION

The most powerful polycomponent techniques are the solution growth methods. Such methods are either isothermal where supersaturation is produced by:

1. A temperature differential (generally nutrient is dissolved in a hot zone and growth is in a cool zone).
2. Solvent evaporation.
3. Mixing reactants

or non-isothermal, where supersaturation is generally produced by:

1. Programmed cooling.

The methods may also be characterized by the nature of the solvent used:

1. Water at ambient or near ambient conditions.
2. Water (plus complexing agents or mineralizers) at hydrothermal conditions.
3. Molten salt solvents.
4. Other solvents including organics, liquid metals and inorganic solvents.

The two most used techniques employed for aqueous solution growth are the rotary crystallizer and the mason jar method. In the rotary crystallizer, seeds are mounted on a reciprocating mechanism which rotates them clockwise for several revolutions and then counterclockwise. Supersaturation is produced by programmed cooling or solvent evaporation. Good control with large crystals is not difficult to achieve.

The mason jar method is simple and inexpensive but control is poor. Seeds are introduced into a saturated solution in a sealed jar which is allowed to cool to room temperature or allowed to isothermally evaporate. Growth of water soluble crystals on a large commercial scale by the temperature differential method has been practical in several instances. Growth by reaction where reactants are mixed almost always results in such high local supersaturation that growth cannot be controlled. The most notable exception is gel growth where the reactants diffuse toward one another in a controlled manner through a gel medium. Growth by electrochemical reaction has been neglected but should have considerable usefulness.

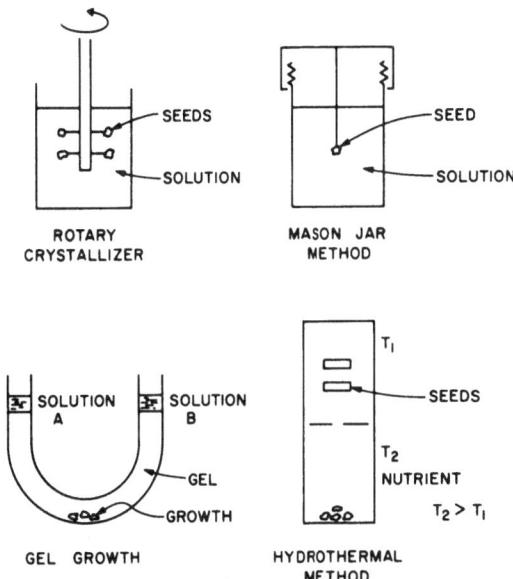

FIGURE 3. Solution Growth Methods

If a material is insoluble in water at ambient conditions, often it has appreciable solubility at elevated temperatures and pressures (hydrothermal conditions). Supersaturation is generally produced by a temperature differential and circulation is aided by the fact that $(\frac{\partial \rho}{\partial T})_p$ (the temperature coefficient of density at constant pressure) for hydrothermal solution is quite large. Solubility is usually increased by the addition of a complexing agent (its role is analogous to the complexing agents used to increase volatility in vapor growth) usually called a mineralizer. Hydroxide is the most common mineralizer. Low temperature polymorphs such as α-quartz are among the materials to which the hydrothermal technique is especially applicable. The use of mineralizers at ambient conditions deserves more investigation. Se has been grown from $S^=$ solutions at ambient conditions.

Molten salts are powerful solvents and have been much used for inorganic salts and oxides. Because of the high viscosity of the medium, diffusion problems, dendritic growth and flux inclusions are often problems in this technique. Usually supersaturation is produced by slow cooling in the absence of seeds, although some experiments with temperature differential growth on seeds show promise.

Perhaps the best known solution growth experiments from other solvents are those involving the growth of GaP from molten Ga.

Solution growth and vapor growth have been combined in the vapor-liquid-solid (V-L-S) method where vapor transport brings material to a liquid drop of solvent in which it dissolves. The material then moves through the drop to a cooler region where growth takes place. The usual morphology of V-L-S grown crystals is whiskers. So far, growth of large crystals has proved difficult.

REFERENCES

Solid-Solid Transformations

1. W.C. BURGERS and K.T. AUST, The Art and Science of Growing Crystals, ed. J.J. Gilman (1963)
2. R.A. LAUDISE, J. Phys. Chem. Solids, Supplement 3 (1967)
3. M.V. KLASSEN-NEKLYUDOVA, Mechanical Twinning of Crystals (1964)

Pure Melt-Solid

1. B. CHALMERS, Principles of Solidification, John Wiley, New York (1964)
2. H.C. GATOS, Properties of Elemental and Compound Semiconductors, Vol. 5, Interscience, New York (1960)
3. D.T.J. HURLE, Mechanisms of Growth of Metal Single Crystals from the Melt, Pergamon, New York (1962)
4. K. NASSAU, J. Phys. Chem. Solids, 23, 1511 (1963)

Vapor Growth

1. J.P. HIRTH and G.M. POUND, Condensation and Evaporation, MacMillan, New York (1963)
2. J.R. O'CONNOR and J. SMILTENS, Silicon-Carbide — A High Temperature Semiconductor, Pergamon, New York (1960)
3. PHILIP S. SCHAFFER, J. Am. Ceram. Soc., 48, 508 (1965)

Solution Growth

1. J.J. GILMAN, The Art and Science of Growing Crystals, John Wiley, New York (1963)

General

1. K. LARK-HOROVITZ and VIVIAN A. JOHNSON, Solid State Physics, Vol. 6, Academic Press, New York (1959)
2. W.D. LAWSON and S. NIELSEN, Preparation of Single Crystals, Butterworths, London (1958)
3. J.W. MULLIN, Crystallization, Butterworths, London, (1961)
4. R.G. RHODES, Imperfections and Active Centres in Semiconductors, MacMillan, New York (1964)
5. A. SMAKULA, Einkristalle, Springer-Verlag, Berlin, Göttingen, Heidelberg (1962)

7/ The Empirical Approach to Superconductivity

B. T. MATTHIAS

I. INTRODUCTION

It is interesting to consider some of the reasons why there is comparatively little discussion these days about new superconductors. First, super-conductivity is a cooperative phenomenon for which no quantitative answers exist. Cooperative phenomena are phenomena that occur only if there is a large number of individuals participating. Such phenomena are not confined to physics; they are a major occurrence in nature. Superconductivity is a cooperative phenomenon of electrical conductivity, but as long as one tries to explain superconductivity as some kind of pathological electrical conductivity, there is not much hope of predicting new superconducting materials. Electrical conductivity is not dependent upon a cooperative phenomenon; it is abundant everywhere. Every metal has electrical conductivity, and even every insulator has some electrical conductivity. Yet superconductivity occurs only under certain conditions involving not only the conduction electrons but mostly the valence electrons.

As a cooperative phenomenon, superconductivity has to be considered in a generous way without concentrating on one or the other specific detail. One must not limit himself to a careful and self-consistent d e s c r i p t i o n of superconductivity, for he will then get a well-developed and perfect picture of the superconducting s t a t e, but he will never get any predictions as to the occurrence of superconductivity.

II. THEORY AND SUPERCONDUCTIVITY

Most theories of superconductivity, in my opinion, are merely descriptions. They become theories only when they are able to predict. Since after 17 years we haven't gotten any transition temperatures, higher or otherwise, from the theoretical point of view, I would say the time has come to consider other ways. I have always thought that the empirical approach is a much more successful one as far as the problems of the discovery and the raising of superconducting transition temperatures are concerned.

Research at La Jolla sponsored by the U.S. Air Force Office of Scientific Research, Office of Aerospace Research, U.S. Air Force, Grant No. AF-AFOSR-631-67.

At present, in all theories, the one crucial parameter is the transition temperature. In other words, unless there is a transition temperature to begin with not much can be gained. It has always been my attempt to find the transition temperature first of all, since then the fact exists. Today, the variation of the transition temperature is more or less understood, but — let me reiterate — strictly on an empirical basis. This may sound sort of pessimistic, but it really isn't as bad as all that. The difficulty with predicting transitions in cooperative phenomena is not restricted just to superconductivity. Let me mention some other cooperative phenomena which are, possibly, in worse shape than superconductivity.

III. OTHER COOPERATIVE PHENOMENA

One familiar cooperative phenomenon is, of course, the melting process. Yet, still today there isn't really a trace of a theory for the melting point. Needless to say, again nobody can predict any melting temperature, and the whole concept of melting is very much ignored today — perhaps because it is such an embarassing situation not to have the slightest idea what might determine the melting point. I will discuss this later, since I have come to the conclusion that, in order to get a somewhat more general idea about superconductivity, one has to be willing to look at other cooperative phenomena in just such a general way.

Besides melting, some other cooperative phenomena are, for instance, ferroelectricity, which is probably better understood than any of them, and ferromagnetism, which is even less well understood than melting or superconductivity — mostly for the reason that there are very few ferromagnets in this world. You see, the more occurrences one has of a phenomenon, the better it can be understood. The exception is the melting point, since everything melts in one fashion or another and yet there still isn't much of an understanding of the process involved.

From this point of view, it seems like a good idea, in order to get a picture of the empirical nature of the superconducting transition temperature, not to look at all at the electric resistivity. For instance, if you wish to look at the resistivity, you either measure it yourself — and you most likely won't have a sample which is clean enough — or else you take data from the literature. Whichever you do, the data can be made to fit wherever necessary because both sources can be so varied.

So what should be done? After what I have mentioned about cooperative phenomena, it is wrong to look at just one electron — one must look at all the electrons which are active in one way or another. Thus in a metal or an intermetallic compound, one must consider all those electrons which are outside of filled shells, since once in a filled shell they will obviously no longer contribute very much. By considering all electrons outside filled shells, one suddenly gets a very clear picture for the behavior of the superconducting transition temperature. Mind you, it's all empirical; however, it works well and it is quite useful in predicting high transition temperatures. As a matter of fact, sometimes the predictions are better than nature can produce. It often happens that a predicted compound will not exist probably because its transition temperature would have been too high. It appears then that those metals which would have had very high predicted transition temperatures actually turn out to be crystallographically unstable.

Let us define the electron concentration in a completely unambiguous and non-artificial way — that is, we shall not use Hall data, resistivity data, or any arbitrary definitions of electron concentrations, but rather just the periodic system. We start with the alkalis and count the valence electrons as one, and we arrive at platinum where we shall count the valence electrons as ten. It is that simple—there are no definitions, no agreements, no conventions on how to count them. If we adhere to this, we then get a very clear-cut picture for the behavior of the superconducting transition temperature, and I will show you in a little while how it looks.

Now let me briefly discuss the other cooperative phenomena which I have mentioned before. Ferromagnetism and ferroelectricity are cooperative phenomena which usually occur at intermediate temperatures. The melting point, of course, by definition always occurs at a comparatively high temperature; and superconductivity, as far as I am concerned, always occurs at a very low temperature. Yet, as you will see later on, all these phenomena are quite intimately related. It is hard to believe, I realize, that superconducting transitions and melting points have anything in common, but once we really started to look for it, we found it. As a matter of fact we found much more. We found mistakes in the periodic system — small ones, but nevertheless mistakes. We also found new relationships in the periodic system, and all this by looking at superconductivity in a purely empirical way. Also — for the first time, I think — we have today the germ of an idea suggesting what the melting point may be due to, and this is the part of the talk I would like to emphasize more strongly than the rest, because I have been very much intrigued during the last half year by the fact that the melting point and the superconducting transition are so closely connected.

IV. MELTING POINTS

One of the most frequently voiced fallacies is the opinion that melting point and bonding strength are synonymous. It is crucial that one realizes that the bonding strength has rather little to do with the melting point in itself. The bonding strength is primarily indicated by one thing only — namely, the boiling point. After all, by definition, when a bond breaks entirely the metal will evaporate. Thus bonding strength is only of secondary importance for the melting point. What really does happen at the melting point is that a certain electronic configuration becomes more spherical or symmetric or isotropic. In other words, the transverse mode of the lattice vibration disappears. This seems to me the only meaningful definition of the melting point. The transition energies involved in the melting process are much larger than those of superconductivity at its transition point, yet they are small compared to those at the boiling point. Superconductivity involves all the electrons outside filled shells, and it is the steric rearrangement of these same electrons which determines the melting point. Thus, if the situation is favorable for superconductivity, one can see immediately that there will be a definite connection between the melting point and the superconducting transition temperature.

Throughout the periodic system, superconductivity occurs in two groups which are separated by non-superconducting elements (See Figure 1). Non-superconducting means only that no superconductivity has yet been found above the millidegree temperature range. It is possible that in the future some of the elements bordering these groups will also become superconducting once higher purities and lower temperatures have been achieved. The exceptions will be the magnetic elements, which occur again in three distinct groups. Only one element, uranium, seems to be an exception to this overall categorization.

Uranium is one of the most fascinating of elements. In one form — this is my opinion — it is antiferromagnetic. In other forms, that is, in other crystal structures, it is superconducting. Uranium also becomes superconducting in the form referred to first (α-U), but only under pressure. This form is the orthorhombic modification of uranium and the one stable at room temperature. The other two modifications of uranium are just superconducting and not magnetic. I shall say more about uranium below, since by studying its superconductivity we have recently learned quite a bit about this metal.

At the lower part of the periodic system are the rare earths, which begin with lanthanum and end with ytterbium. Here again the interplay between superconductivity and magnetism is evidenced. I would rather not talk about those elements not showing either phenomenon at this time because things are presently too much in flux regarding their properties and have not yet become clear. We have made a change in the periodic system — a result that emerged from the superconducting behavior. Below scandium and yttrium, in days gone by, there was lanthanum. We changed it. We replaced lanthanum by lutetium and you will see shortly why this area of the periodic system had to be changed. In the periodic system most properties are periodic. The fact that only lanthanum was superconducting, while scandium, yttrium, and lutetium were not, was rather confusing — to say the least — until we suddenly realized that maybe the periodic system was wrong in this respect and maybe lutetium should have been below scandium and yttrium instead of lanthanum. We changed it and were rather proud of ourselves until we found one day that Landau and Lifshitz in their book ten years ago had already stated precisely this fact — that there was a mistake. We really had been unaware of this, but were reassured when we found out.

V. MECHANISMS OF SUPERCONDUCTIVITY

It has been my opinion for a long time that the mechanism which causes superconductivity is frequently not just an electron-phonon interaction. Undoubtedly, among the non-transition elements (see Figure 1), the interaction between the electrons and the lattice vibrations does cause superconductivity. But I always suspected that there had to be other mechanisms too. First, one has two distinctly different groups of superconductors. Now, among the ferromagnets, ferromagnetism arises through many different mechanisms. Why, therefore, should we limit ourselves to one mechanism only for the superconducting metals. Among the transition elements, I was always convinced, there had to be a different mechanism, and recently we have come to the conclusion that again for protactinium, uranium, and lanthanum there may be yet another, still somewhat different mechanism responsible for their superconductivity.

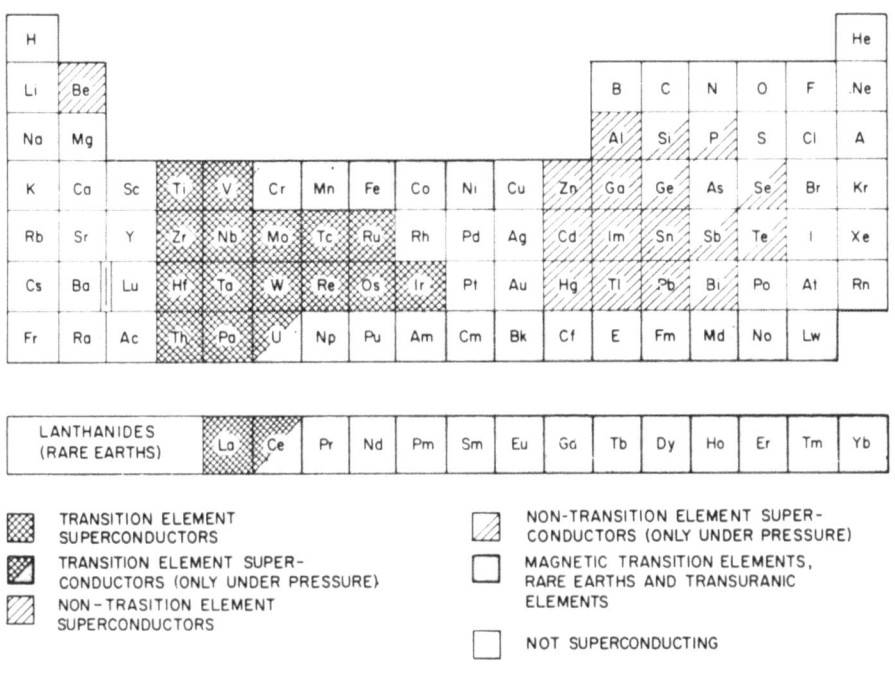

FIGURE 1

For the non-transition elements the interactions between electrons and lattice vibrations are perfectly straightforward, since one invariably finds for these elements an isotope effect of the form

$$T_c \sim \frac{1}{M^{1/2}} \,.$$

That is to say, the transition temperature changes with the isotopic mass according to the law of frequency vs. mass of a harmonic oscillator. For a long time we had tried to show that for the transition elements the situation was different. In order to prove this, since nobody was willing to believe it, we decided to measure the isotope effect for the transition elements. Today, isotope effects have been measured for many of them and there has not yet been a single transition element for which the isotope effect is $T_c \sim 1/M^{1/2}$. For Zr and Ru the mass exponent is zero; for molybdenum it is $1/M^{1/3}$; for tungsten perhaps it might be $1/M^{1/2}$, but we don't know yet. To date anyway, the isotope effects among the transition elements are always quite different, indicating that the mechanism must be different.

Now, why are we so interested in this? Most non-transition elements become superconducting, but the transition temperatures never exceed 10°K. In the transition elements, whenever they become superconducting, they really do it with a bang. The transition temperature in relation to the electron concentration is extremely sensitive to the latter, as shown in Figure 2. In this figure, the total number of points is close to 500, including elements, compounds, and alloys. The very sharp dip in the middle is probably an indication of the enhanced stability of the half-filled d-shell. While a half-filled shell is not as inert as a filled shell, it nevertheless approximates it somewhat. There have always been three exceptions to this plot: lanthanum, protactinium and uranium. In the past, uranium, with six valence electrons, has erroneously been accorded T_c's ranging from ~ 0.7-$1.7°K$. More recently, protactinium, with five electrons, is a third exception. It is superconducting at $1.7°K$. We now have reason, however, to think that these three metals are not really exceptions, but rather constitute a third group of superconductors. So you see, with the transition metals, if one adheres to electron concentrations near five or seven, one will usually achieve very high transition temperatures, provided the crystal structure is correct.

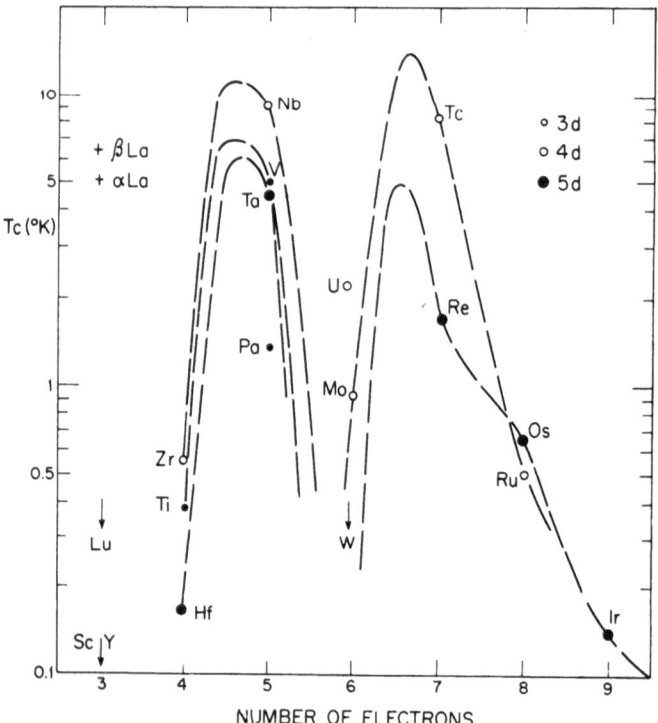

FIGURE 2

In addition to the interaction of the electrons with the lattice vibrations, what is the other mechanism? In the first mechanism, one deforms the lattice but treats the atoms as rigid spheres. But who really believes this completely? Looking at the electron shell, one can hardly consider it a rigid sphere. The electron shell in itself must also be deformed or polarized; and, considering this as another interaction mechanism, one realizes that an increase in atomic mass will give the opposite effect on the interaction as compared to the lattice vibration. For conduction electrons to interact with lattice vibrations, the more the lattice vibrates the better. Thus lower isotopic masses will give higher transition temperatures. However, if it is advantageous to polarize an atom — its electron shell, that is — one will get just the opposite mass dependence, because one can polarize more easily the larger the mass. If the mass is very low, distortion of the lattice will take place, but not polarization of the atom. So here is the difference — it's not so strange any longer — between the superconductivity of the transition elements and that of the non-transition elements, and, in my opinion, it is the reason for the difference in the isotope effects as well as for the very high transition temperatures which we can get only with the transition elements and their compounds.

Before going into more detailed results, let me mention again why I consider lanthanum, protactinium, and uranium to be different. First of all, they don't fit the pattern of superconductivity evidenced in Figure 2. Now, in the occurrence of superconductivity throughout the periodic system, lanthanum, protactinium, and uranium have something else in common: right after them the f-electron series begin. Following lanthanum in the lanthanides, as is well known, magnetism begins with cerium — a consequence of the localized $4f$-electrons of the rare earths. The same thing happens again with protactinium. From protactinium on, $5f$-electrons will occur. Therefore the term "thorides" would be more correct and representative of this group than "actinides." There never have been f-electrons in thorium and there never will be, neither in thorium nor in any of its compounds. In agreement with Zachariasen, I think this grouping should be called the "thoride" series, or, if you wish, a "protactinide" series, since that is where it really begins. But the final evidence for this was discovered only recently — again by considering the superconducting behavior — even though Zachariasen had stated it a long time ago.

Uranium is another example for the new superconducting metallurgy. If one has alpha uranium, i.e., the uranium modification stable at room temperature and below, it is, all by itself, not superconducting. For many years, people recorded the superconductivity of α-uranium, but there never was any. What had been observed were small filaments of something that was superconducting. The bulk of the sample never became superconducting. Well, at first it was assumed that uranium must be a gapless superconductor — in other words, a theoretical term describing the fact that one didn't really understand what was happening. Now, however, there is nothing gapless whatsoever. The truth is that in spite of all our knowledge of uranium, nobody ever had any very clean uranium. As we got cleaner and cleaner uranium, in fact, its superconducting properties began to disappear entirely. What was it then that people had seen previously? The answer was forthcoming — it takes very little pressure to render uranium superconducting. The cause is a change in the f-electron configuration. At normal pressure, uranium becomes, in all likelihood, antiferromagnetic (at very low temperatures — below 0.1°K). Under hydrostatic pressure one changes the f-electron configuration somewhat and immediately superconductivity appears. And this, in my opinion, is one of the really intriguing results of the study of superconductivity: to realize what a complicated situation exists for uranium — namely, one doesn't know precisely when the f-electrons will come into play. Subsequently, measurements of the specific heat of uranium showed that under pressure uranium turns into a good, clean-cut superconductor. Without any pressure there is an indication of a magnetic transition — again through the specific heat. So this may be the explanation of why uranium takes a sort of in-between or Zwitter position in the periodic system.

VI. OTHER PROPERTIES

The melting point is not the only macroscopic property to which superconductivity is related. Hardness, for instance, is another. It is intriguing to realize that the composition of the superconductor with the highest known T_c [$Nb_3(Al_{0.8}, Ge_{0.2})$, $T_c = 20.5°K$] coincides with the composition of minimum hardness in the pseudo-ternary system

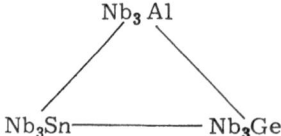

Whether this is indicative for melting temperatures cannot be decided at such an early stage.

The encouraging part of this is that, by looking at these macroscopic properties, one can find new superconductors and, as a matter of fact, destroy old superstitions — such as the one I had for many years that 18°K was the upper limit of superconducting transition temperatures and that there was just no way to get above 18°K. Well, the new result is only a few months old and one might be cheered by the fact that this was possible after so many years. Let me point out again: in order to get to higher temperatures in superconductivity, it is absolutely necessary to look in a more general and less limited way from every point of view as far as macroscopic properties are concerned.

Let us illustrate this by the third column of the periodic table. It consists of scandium, yttrium, lanthanum, the magnetic rare earths, and lutetium, which is not magnetic. For almost ten years we searched for superconductivity in scandium, yttrium, and lutetium because lanthanum was superconducting. These attempts were entirely unsuccessful until the question was inverted and the problem approached by asking why lanthanum became superconducting while all remaining (non-magnetic) elements in this column did not. Usually, if one only asks the right questions, one has at least half the answer. We suddenly realized lutetium should be in the position of lanthanum in the periodic system and lanthanum itself should be close to barium instead. It took another four or five years for me to realize that still more was strange about lanthanum. Looking at the melting points again, Sc, Y, and Lu all melt between 1400 and 1600°C. Lanthanum, however, melts at slightly above 900°C. Thus lanthanum doesn't fit at all into the periodic behavior of the periodic system. That is, it's about 500 degrees too low for the corresponding group in the periodic system. This behavior of the melting point must, of course, be linked very closely to the mechanism for superconductivity. We soon came to realize that similar behavior (and even more pronounced) would be evident for the $5f$-elements. Beginning with protactinium, the melting point is already about 700 to 800 degrees too low, and this anomalous depression increases to a maximum for neptunium and plutonium.

For the $5f$-series, again protactinium and the element following it, uranium, are superconducting and again they are superconducting in an anomalous way. The electronic configuration responsible for this superconductivity is also the one responsible for the very low melting points.

The superconducting behavior of uranium is one of the most complicated ones known. There are three crystalline modifications of uranium, two stable only at elevated temperatures. If these two modifications are retained at low temperatures by alloying with stabilizing elements they both become superconducting. The modification which is the stable one at low temperatures becomes superconducting only when compressed under hydrostatic pressure, as mentioned before. Thus, the superconductivity that had been ascribed to this form of uranium was due to either impurities or strain and indicative only of the imperfect nature of the uranium investigated. Superconductivity was finally the decisive approach to revealing all these complications.

A similar situation has been discovered recently for the superconducting behavior of beryllium, and indeed, the simultaneous existence of several polymorphic superconducting modifications has become apparent. Beryllium itself isn't superconducting, but the presence of the slightest amount of transition metal impurities will immediately make it superconducting, as well as stabilize other modifications. It takes as little as 10 parts per million. When finally really purified, beryllium becomes as ductile as tin and may superconduct at very low temperatures. [Note added in proof: Falge has recently reported the superconductivity of pure Be at 0.026°K.]

VII. CONCLUSIONS

To conclude, if one has a material which doesn't do what it should from the superconducting point of view, one can assume quite safely that something is wrong with the metallurgy. Adhering to our existing expirical approach (and mind you empiricism means that we don't really understand what we are doing, we simply have a rule to cling to which works), one can find practically an unending number of superconductors.

In order to raise the transition temperature, two things must be observed. First, one must involve not just the lattice deformation potential, but also the deformation of the electron shell. This is absolutely necessary. Otherwise superconductivity will never occur above 10° or 11°K. Secondly — and I don't know why this is so, I just know that it is — one must have a crystal with cubic symmetry. So far, all the systems in which we have found superconductivity above 15°K — and there are quite a number of them — have always been cubic. Once, when we found a high T_c in a hexagonal phase, a closer look showed that we had a cubic component unknown from existing phase diagrams.

So, this is my prescription: adhere to five or seven electrons, involve the electron polarization as compared to only the lattice polarization, and remain in the cubic system. Considering all this, I think it is feasible to increase the transition temperature somewhat above the present 20.5 degrees to near 22°K.

REFERENCES

1. J. J. ENGELHARDT, G. W. WEBB and B. T. MATTHIAS, Science, **155**, 191-193 (1967)
2. R. H. WILLENS, E. BUEHLER and B. T. MATTHIAS, Phys. Rev., **159**, 327-330 (1967)
3. B. T. MATTHIAS, T. H. GEBALLE and B. T. COMPTON, Rev. Mod. Phys., 35 (1963)
4. B. T. MATTHIAS, IBM J. Res. Devl., **6**, 250 (1962)
5. J. P. REMEIKA, T. H. GEBALLE, B. T. MATTHIAS, A. S. COOPER, G. W. HULL and E. M. KELLY, Phys. Letters, **24A**, 565 (1967)
6. A. R. SWEEDLER, CH. J. RAUB and B. T. MATTHIAS, Phys. Letters, **15**, 108 (1965)
7. CH. J. RAUB, A. R. SWEEDLER, M. A. JENSEN, S. BROADSTON and B. T. MATTHIAS, Phys. Rev. Letters, **13**, 746 (1964)
8. G. ARRHENIUS, CH. J. RAUB, D. C. HAMILTON and B. T. MATTHIAS, Phys. Rev. Letters, **11**, 313 (1963)
9. B. T. MATTHIAS, T. H. GEBALLE, K. ANDRES, E. CORENZWIT, G. W. HULL and J. P. MAITA, Science, **159**, 530 (1968)
10. T. H. GEBALLE, B. T. MATTHIAS, J. P. REMEIKA, A. M. CLOGSTON, V. B. COMPTON, J. P. MAITA and H. J. WILLIAMS, Physics, **2**, 293-310 (1966)
11. T. H. GEBALLE, B. T. MATTHIAS, K. ANDRES, E. S. FISHER, T. F. SMITH and W. H. ZACHARIASEN, Science, **152**, 755-757 (1966)
12. B. T. MATTHIAS, Phys. Letters, **25A**, 226 (1967)
13. B. T. MATTHIAS, T. H. GEBALLE, R. H. WILLENS, E. CORENZWIT and G. W. HULL, Jr. Phys. Rev **139**, A1501-A1503 (1965)
14. B. T. MATTHIAS, T. H. GEBALLE, L. D. LONGINOTTI, E. CORENZWIT, G. W. HULL, R. H. WILLENS and J. P. MAITA, Science, **156**, 645-646 (1967)
15. B. T. MATTHIAS, Phys. Rev., **97**, 74 (1955)
16. B. T. MATTHIAS, Ferroelectricity, Ed. F. Weller, p. 176, Elsevier Publishing Company, Amsterdam (1967)
17. C. E. OLSEN, B. T. MATTHIAS and H. H. HILL, Z. Physik, **200**, 7-12 (1967)
18. J. WITTIG and B. T. MATTHIAS, Science, **160**, 994-995 (1968)

8/The Material Aspects of Computor Storage

W. D. DOYLE

Part I

I. INTRODUCTION

Since a considerable portion of the research in solid state physics throughout the world has been motivated by the search for improved computer storage, it is reasonable to occasionally assess progress in this regard. It is my purpose to do so here. To give a feeling for the progress made, Figure 1 shows the original BINAC (1948) and Figure 2, a modern computer system (1967).

FIGURE 1. Computing System, circa 1948. The machine shown is the BINAC, one of the first electronic computers.

Computer storage [1] embraces all aspects of the solid state and it would be unwise to attempt to reduce solid state physics to a short paper. I am afraid that my evaluation of this problem is influenced by my own association with magnetics, although fortunately the dominating role of magnetics cannot be denied. I shall review the requirements for storage, presenting a very simple picture of how a "memory" (which is the storage device) operates and the definition of a few terms which are occasionally required.

I will then go on to the specific phenomena which are used or which may be used for storage. Each of these will be briefly described so that before I discuss any one in detail, the reader will have the whole picture as it is today.

FIGURE 2. Computing System, circa 1968. The machine shown is a Univac 9400.

Magnetic storage, in particular ferrite cores and thin films, will then be discussed. I plan to go more deeply into these areas since magnetic storage, as I mentioned, presently dominates the field and will for many years to come. Thin films have an additional attraction in that they provide an excellent medium for studying magnetic effects.

Finally, I will treat magneto-optic and photochromic memories. They are both candidates to replace cores and films and are rich in unsolved and exciting research problems.

I do not intend to discuss semiconductors in detail, although their impact on the storage field is increasing rapidly. However, since the main problem with semiconductor memories is frabrication and not the material itself, there is some justification in addition to my ignorance for overlooking it here. On the other hand, it is not clear to me that superconducting memories will achieve broad acceptance and so with all due respects to those active in that field, I shall give it only a passing glance.

II. THE CHARACTERISTICS OF A STORAGE DEVICE

A. *The Storage Method*

I would like to describe a very good storage device. Take an electric switch: it has two stable states, goes from one to the other very positively, and stays there forever until I change it. Let me call changing it the "write cycle". It says "on" and "off" so I can tell by looking at it which position it is in. Let me call looking at it the "read cycle". Notice also that when I look at it, it stays where it was. Therefore it exhibits "non-destructive" readout. It requires no power except when I read or write. Furthermore, if the power fails, it stays where it was so it is "non-volatile". Each switch could represent one "bit" of information. This bit is either "on" or "off".

Now the logician uses a number of switches together to represent a more complete piece of information. The scheme used would depend on the phenomena used for the "bit". All practical devices to date have been either binary or binary coded, i. e., "on" or "off", so that is the scheme used. Instead of "on" or "off", we use "1" or "0". For a number like 3, we write "11", for five "101", etc.

B. *The Physical Characteristics of a Storage Element*

The problem with our switch is that it operates slowly, requires me to push it to write and to look at it to read. So before we start building a memory, let's find out what we want. What characteristics should the phenomena have?

1. Two Discrete States (or more). Presently, all memories are binary but logic could be developed for multiple state logic.

2. Threshold. We would like the device to switch sharply from one state to another whenever a critical value of some parameter is exceeded.

3. Stable. We would like the device to stay in its determined state in the absence of applied power (if possible). There are two modifications of this: a) The device sits at rest in its state until we decide to change it. b) It delays for some fixed time a signal which we can circulate around in a closed loop.

We also would like the device to be temperature stable and time stable.

4. Writeable. We want to be able to change the state of the device easily.

5. Readable. We want to be able to determine the state of the device easily, hopefully without disturbing it.

C. *Practical Characteristics*

However intriguing a particular phenomenon may be to the researcher, its value is determined finally in the market place. The criterion there is c o s t. Therefore, there are practical characteristics which the storage element must possess.

1. H i g h   S p e e d . Since one pays for time used on a computer, the faster it can operate at a given cost, the more economical it will be. The single most important factor limiting a computer's overall speed is the main memory cycle time.

2. L o w   C o s t. In addition to speed, another equally important characteristic is the cost, usually quoted as the cost/bit. Unless this number is competitive with other techniques, sizeable improvements in performance would be required before any interest would be aroused.

I apologize for sounding like a salesman, but there is an important point here which researchers often overlook. Devices which are too expensive to make will not be made. However beautiful the research, cost will always haunt the researcher and in the end will determine if his work leads to a practical device.

3. S t o r a g e   C a p a c i t y. As a general rule, the bigger the memory is (i. e., the number of bits), the better. Obviously, the bits must be physically small and amenable to close packing if the memory is to have reasonable dimensions and maintain its speed.

III.   STORAGE PHENOMENA

I don't think it will come as much of a shock to say in advance that to date we have not found the perfect element. In general, fast memories have been small, and big memories have been slow. Fast memories have been expensive and slow memories have been cheap. The situation now is that memory systems are broken down into functions so that in a particular computer one may find three different kinds of memories:
a) large, slow storage, b) medium speed, medium size storage, and c) small, fast storage.

As I go along, I shall point out which phenomena are suitable in the various capacities.

A. *Magnetics*

It is a general characteristic of magnetic materials that a plot of M vs. H will be hysteretic. By proper treatment, it is possible to prepare a large variety of materials in various shapes which possess very square loops.

Consider an ideal loop first. It fits very well the criterion we have established for a storage element. How fast is it? The answer to that will occupy a fair portion of the paper. It is enough to say that it is ~a microsecond or less, which is sufficient to make it a reasonable element.

Probably the most common magnetic memory is magnetic tape (not for high-speed storage but for slow, mass storage). Here a slurry of fine magnetic needles is smeared on a plastic tape. The particles are aligned along the tape and can be magnetized either parallel or antiparallel to the tape length to represent a one or zero. A magneto-mechanical head is used to both produce the write field and also to detect the stored information non-destructively.

Now a large block of magnetic material generally does not have a square loop. This occurs because of the occurrence of domains. However, domain walls have a certain energy and, therefore, as the size of the particle decreases, the energy balance swings in favor of the single domain state and the particle has a very square loop. This size will depend on the magnetization of the material. Most tapes are made from iron oxide [Fe_3O_4 (magnetite) or Fe_2O_3] which has single domain size of ~1000Å.

For single domain particles, the coercive force will depend on the shape so that one can tailor the length/width ratio of the particle to obtain the appropriate coercive force. For tapes, H_c ~ 25 oe. The advantage of tape is that it will hold a lot of information and is very cheap. Typical characteristics are 1600 bits/in., 7-9 tracks on $\frac{1}{2}$ in. tape, 2400 ft./rolls moving at ~10 ft./sec.

A considerable improvement on tape has been the magnetic drums and disks. Both utilize a plated alloy, often NiFeCo, with a coercive force of 40 oe. Considerable research has been directed towards finding the reason for the high coercive force. It is now thought that the plating produces not a film but an agglomeration of single domain size particles.

Drums commonly have capacities of 10^7-10^8 bits. They rotate at speeds from 700-900 rpm. Normal values of 50 tracks/in. with 1000 bits/in. Disks have about the same capacity but are removable. The drum is a huge fixed installation. Typical Fastrand drums in various stages of completion are shown in Figure 3.

Another advantage of drums and disks with respect to tape is that they are semi-random access. I have not discussed the organization of the memory but you might imagine that it would be advantageous to go to any bit directly from any other bit. This is called random access. In a tape, all the bits between the two of interest must also pass by the detector. In the drums and disks the heads move and the drum and disks rotate, so that only some of the intermediate bits pass under the heads.

We now logically come to the work horse of the computer industry, the magnetic core, a pressed ferrite torroid. Because of its importance, I have reserved a separate part for it. It is nothing more than a magnetic doughnut which can be magnetized in one of two directions to represent ones and zeros.

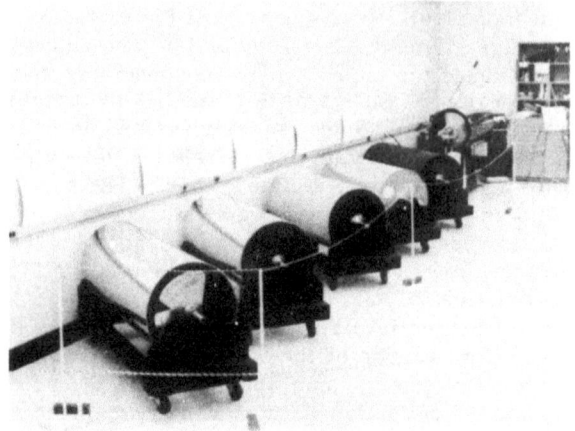

FIGURE 3. Fastrand data storage drums shown in several stages of completion.

The other important magnetic element is the magnetic film which is just now coming to the marketplace. My discussion of that element will, I hope, give a very good insight into the magnetics of solids. It depends for its life on a phenomenon which is to some extent still a mystery. It is possible to make a film in which the magnetization prefers to lie along a unique axis. Thus we can again use the direction of the magnetization (parallel or anti-parallel to this axis) to represent ones and zeros. We shall see more of this.

B. Semiconductors

A very good storage element is the flip-flop. It has been used for years in the part of the machine where the data is processed, the register. However, to make inexpensive arrays the flip-flop must be small and have low ambient power consumption. This is now feasible with integrated circuits and particularly with metal oxide semiconductors (MOS).

I would like to briefly discuss one type of MOS which is presently of great interest, the silicon-on-sapphire approach [2]. A p or n type silicon single crystal film is grown epitaxially on sapphire etched into discrete islands, then two opposite polarity regions are created by diffusion of boron for p type or phosphorus for n type, an oxide film grown on the silicon and a metal gate added. The conduction is controlled by the voltage on the gate which is really capacitively coupled through the insulator to the device.

The voltage on the gate can invert the layer underneath it. That is if it was originally n type, then a negative voltage will repel the electrons, attract the holes and create a p type connecting the p type source and drain.

The actual mechanism in the MOS is more complicated but is outside the scope of this lecture. Germanium has a higher mobility than silicon, but it is difficult to grow the oxide so that silicon dominates MOS technology.

In the complimentary MOS flip-flop, in which two n and two p types are used, it is possible to reduce the ambient power to a very low level.

This eliminates one of the serious drawbacks of semiconductors which are active storage elements. Furthermore, the entire cell occupies a space less than 15×15 mils. MOS devices of 10^4 bits are estimated to operate at < 100 ns. The very big unanswered question is whether large arrays can be made economically. Since the critical voltages required to switch depend on film thickness etc., it must be possible to produce uniform arrays. So far, this success hasn't been assured. (Recently, IBM announced a 10^5 bit, $80\mu s$ memory.)

C. *Superconductors*

It is well known that the resistance of many materials changes dramatically from a finite value to a very small value (or zero) below a critical temperature T_c near absolute zero. In addition T_c is dependent on magnetic field. These facts have been utilized to make, at least scientifically, a very exciting storage device [3]. By using material with two different values of H_c, we can change one from superconducting to non-superconducting using the field from a current in the other. If the current carrier has the higher H_c, it will stay superconducting. Evaporated thin films on superconducting substrates are used to confine the flux and reduce the inductance so that the current may dissipate quickly and increase the speed of operation.

The switching time in actual thin film devices is of the order of 100 ns. Since superconducting storage can only be economical if the capacity $>10^7$ bits (because of the cost of cryostat), the cycle time will probably increase to be on the order of a microsecond.

However, to fabricate large planes which have uniform critical currents is extremely difficult and the most severe problem remaining is developing a practical storage device. Most of the early interest in superconducting memories has faded away.

D. *Acoustical Storage*

It may be a surprise to know that the storage system in the BINAC, the first electronic computer, was an acoustic delay line. It was a large tank of Hg with transducers on the ends. By repetitively pulsing or not pulsing the transducers, a string of ones and zeros could be stored for a short time in the Hg. When they got to the other end, they could be stored again by sending them through a conventional electrical circuit array to the front of the delay line. Figure 4 shows one of the cells with its inventor, J. Presper Eckert. It had a cycle time of $1\mu s$.

The acoustic memory was completely replaced by the core. However, recently very good transducers have been prepared using for example, CdS films. Now one can envision a quartz rod with a large number of these transducers on the face and the acoustic memory is again a real possibility.

Obviously, acoustic memories are not random access, but are so called bit-serial. For reasons associated with the cost and type of the logic envisioned in future machines however, it is possible that bit-serial organization would be practical. This has inspired a re-interest in acoustic delay lines in several laboratories.

FIGURE 4. An example of the mercury delay line memory used in the BINAC. Shown with the device is its inventor J. Presper Eckert.

E. Optical Memories

All of the devices discussed required electrical signals to the elements to write and read. For large scale memories, the lines to and from the element result in long delays. Additionally, they must be evaporated in many instances which complicates the construction and increases the cost. Consequently, it would be nice if one could replace the conductors with light beams.

At this time, no operational memories exist which use light beams for addressing the element. However, at least two have been proposed and one is being investigated intensively. I would like to describe them briefly. Later I shall discuss them more carefully.

1. Magneto-Optic Storage. If plane polarized light is incident on a magnetic material then for certain orientations of the magnetization the plane of polarization of both the reflected (Kerr Effect) and transmitted (Faraday Effect) wave can be rotated. The Kerr Effect has been used experimentally to read out tapes and thin films. A very interesting class of materials are the garnets which have very large Faraday constants. Since the Faraday rotation is dependent on magnetization direction, it is possible to store information by letting the

direction of the magnetization represent ones or zeros. A light beam can then be used to read. In addition, the garnets have certain other properties which make it possible to also write with the light beam. One can conceive of storing 10^6 bits on a one inch platelet. The ultimate speed is a matter of discussion but one can expect at least $1\mu s$. I shall discuss this device in some detail below.

2. Photo-Chromic Memories. All of the elements described have utilized the macroscopic properties of a material. (We shall see later than even in the magneto-optic memory, we are obliged to use a macroscopic region.) We needed a chunk of a tape, or core, or semi-conductor, etc. Even though we can at times make them very small, on an atomic scale they are always gigantic. It is the dream of all researchers in the field to some day utilize an atomic effect. Imagine 10^{23} bits if you will!

To date, it has only been a dream. However, recently a suggestion has been made to use the F-color centers in alkali halide crystals. It is an exciting if complicated idea and I shall save it for the end.

F. The State of the Art

To finish this introduction to storage devices, I would like to show a table (Table I) prepared by Hobb [4] in 1966. It is a run-down on what could be done then and it is essentially correct today (April, 1968).

TABLE I.   Summary of the State of the Art for Storage Devices (After Hobbs, 1966)

CATEGORY	PRIMARY TECHNOLOGIES	TYPE OF ACCESS	CYCLE ACCESS TIME	CAPACITY IN BITS	COST PER BIT
DISCRETE BIT STORAGE AND REGISTERS	INTEGRATED CIRCUIT FLIP-FLOP	RANDOM	50 NS TO 500 NS	1 TO 1000	$1 TO $10
HIGH SPEED CONTROL AND SCRATCH-PAD MEMORIES	MAGNETIC THIN FILM	RANDOM	100 NS TO 500 NS	2500 TO 200,000	$0.50 TO $2.00
MAIN HIGH SPEED INTERNAL MEMORIES	MAGNETIC CORE	RANDOM	$0.7\mu S$ TO $4\mu S$	10,000 TO 2,000,000	5¢ TO 25¢
ON-LINE AUXILIARY STORAGE (SOLID-STATE)	MAGNETIC CORE	RANDOM	$3\mu S$ TO $12\mu S$	1×10^6 TO 100×10^6	1.5¢ TO 3.5¢
ON-LINE AUXILIARY STORAGE (ELECTROMECHANICAL)	MAGNETIC DISK FILE	SEMI-RANDOM	15 MS TO 150 MS	20×10^6 TO 2000×10^6	0.01¢ TO 0.2¢
OFF-LINE AUXILIARY STORAGE	MAGNETIC TAPE	SERIAL	–	–	–

Part II

I. MAGNETIC STORAGE - SOME FUNDAMENTALS

In order that we can deal with the problem of magnetic storage in a little depth, I would like to review some of the characteristics of magnetic materials [1-3].

A. Spontaneous Magnetization

The novelty of magnetic material is that in the absence of an applied field, a sample may exhibit a spontaneous magnetic moment. This occurs because the spins interact in such a way that the total energy of the system is dependent on the alignment of the spins. This is the so called exchange energy which we can express as

$$E_x = -2J \sum Si \cdot Sj \tag{1}$$

where J is the exchange integral and S is the spin angular momentum. Ferromagnetic materials have positive values of J and therefore the spins in ferromagnetic materials try to stay parallel. But J can be negative in some materials and in these the spins try to stay anti-parallel. If the spins on adjacent lattice sites have equal value, then the net moment will be zero. These materials are called anti-ferromagnetic. More commonly, the spins have different values so that a net moment results. These materials are called ferrimagnetic.

The common ferromagnetic materials are the metals Fe, Ni, Co and their alloys. Because of the spin alignment, they have high values of magnetization. Typically for Fe, M_s = 1700 emu/cc or 2.22 Bohr magnetons per magnetic atom. Now $\mu = + g\mu_B$ J $\sim + 2J\mu_B$ $+ 2S\mu_B$. Since Fe has six electrons in the unfilled d shell, it has four unpaired electrons or S = 2. Therefore we would expect μ = 4. The fact that μ is much less is due to several things, e.g., spin orbit interaction.

The common ferrimagnetic materials are non-conducting oxides like Fe_3O_4 which are called ferrites. Their structure is much more complicated than the metals but they are often easier to understand. For example, Fe_3O_4 is an inverse spinel with a value of $\mu = 4.1\mu_B$. Although their structure is complex, the fact that ferrites are insulators makes their ferromagnetic properties simpler to understand. In the spinel structure, there are two different sites for the metal ions: tetrahedral sites and octahedral sites. In $Fe_3O_4 = (Fe\ 0 \cdot Fe_2O_3)$ the eight tetrahedral sites are occupied by eight Fe^{+++} ions. The 16 octahedral sites are occupied by eight Fe^{+++} and eight Fe^{++} ions all parallel to each other but antiparallel to ions in the tetrahedral sites. Thus all the Fe^{+++} cancel and we are left with the Fe^{++}. The Fe^{++} have $S = 2$ and therefore we expect $\mu = 4\mu_B$. Thus, for the ferrimagnetic material we can calculate μ simply. In addition to making them easier to understand, the fact that ferrites are insulators has very important device implications. We shall see this shortly.

B. Magnetic Energies

The energy of a magnetic material is dependent on the position of magnetization through a number of mechanisms. Before we proceed, it will be convenient to review them.

(1) The Zeeman Energy Density. $E = -\overline{Ms} \cdot \overline{H}$ (2)

The magnetization tries to align itself parallel to an external field. For Fe, $M_s = 1700$, so that at $H = 1000$ oe, $E_H \sim 2 \times 10^6$ ergs/cc.

(2) The Exchange Energy Per Pair of Spins.
$(E_x)ij = -2J\,S^2\,\cos \phi\,ij$ (3)

The exchange Energy Density can be obtained from Eq. (3) and is

$$E_x = + A\,[(\nabla \alpha_1)^2 + (\nabla \alpha_2)^2 + (\nabla \alpha_3)^2]$$ (4)

Here $A = 2\,J\,S^2/a$ for a bcc lattice where $\underline{a}$ is the lattice constant and α's are the direction cosines of the magnetization. $A \sim 1.7 \times 10^{-8}$ ergs/cm for Fe. The exchange energy tries to smooth out local variations in the magnetization. If all the spins in a sample of Fe deviated by ~ 0.5 degrees from each other then the energy expended would be about 10^6 ergs/cc.

(3) Magnetostatic Energy Density. $E_M = -\frac{1}{2}\overline{M} \cdot \overline{H}_D$ (5)

Here H_D is the demagnetizing field equal to NM where N is the demagnetizing factor. The demagnetizing field can be calculated from the charge density ρ where $\rho = -(\nabla \cdot Ms)$. For a uniformly magnetized cylinder of Fe, ΔE_M is of the order of 10^6 ergs/cc if the length to diameter is ~ 1. Thus in a magnetic needle, it is difficult to turn the magnetization away from the long axis because of the expenditure of magnetostatic energy.

(4) Crystalline Anisotropy Energy Density. The energy of a magnetic single crystal is in general dependent on the direction of magnetization. The anisotropy energy can be expressed from symmetry conditions as an expansion. For cubic crystals

$$E_K = K_1(\alpha_1^2\,\alpha_2^2 + \alpha_2^2\,\alpha_3^2 + \alpha_1^2\,\alpha_3^2) + K_2(\alpha_1^2\,\alpha_2^2\,\alpha_3^2)\ldots$$ (6)

Here K_1 and K_2 are the fourth and sixth order anisotropy constants and the α's are the direction cosines of the magnetization. Normally, one ignores higher orders in K because within experimental error the available data can be fitted using only two constants. (A measurement has recently been reported for K_3 in Ni.) Odd powers are missing because from physical considerations the energy must be independent of the choice of co-ordinates. The second order term $(\alpha_1^2 + \alpha_2^2 + \alpha_3^2)$ is identically equal to unity.

For Fe, $K_1 \sim + 4 \times 10^5$ ergs/cc and $K_2 \sim + 2 \times 10^5$ ergs/cc. Since K_1 and K_2 are both positive in Fe, the easiest direction is (100) and the hardest is (111). If we look in the (100) plane, $\alpha_3 = 0$, so that $E_K = K_1 \alpha_1^2 \alpha_2^2$

$$= K_1 \cos^2 \theta \cos^2 (90 - \theta) = \frac{K_1}{4} \sin^2 2\theta$$

In the permalloy alloys (80% Ni - 20% Fe) the crystalline anisotropy is very small. In fact at 76% Ni, $K_3 = 0$. This is important technologically. The origin of anisotropy is still not well understood and several mechanisms have been proposed. It has been suggested that the spins see the lattice through spin — orbit coupling and the orbital interaction with the lattice is anisotropic.

(5) Stress Anisotropy Energy Density. The magnetic energy of a crystal is stress dependent. This is not surprising if one accepts the existence of intrinsic crystalline anisotropy. Obviously, the crystalline energy will be sensitive to any distortion in the lattice and is linked to the strain through the magneto-elastic coupling constants. Conversely, changes in the magnetization will cause changes in the strain.

Unfortunately, magneto-elastic phenomena are not simple and I do not think I can coherently describe them here. As a short cut, let me say that one normally considers two magnetostriction constants λ_{100} and λ_{111} which are related to the coupling constants through the elastic constants. For isotropic material, we can say that $\lambda_{100} = \lambda_{111}$ and where T is the tension and

$$E_\sigma = 3/2 \, \lambda \, T \sin^2 \theta \tag{7}$$

where θ is the angle between the magnetization and the tension. Thus tension (or compression) will lead to uniaxial anisotropy. For Fe, $\lambda \sim -10^{-5}$ so that a stress of $\sim 10^9$ dynes/cm^2, causes an anisotropy $\sim 10^4$ ergs/cc.

The importance of stress effects cannot be overemphasized. Because they are awkward to handle, they are often badly treated. We shall see what a considerable role they play in the permalloy alloys.

(6) Special Anisotropies. In the investigation of magnetic material, effects have been observed which are unique to particular systems. One of these is ion-pair ordering. Since it will dominate our development of iron-nickel films, I will postpone discussion of it until later.

C. Magnetic Domains

One of the first mysteries of ferromagnetization was the "now you see it, now you don't" aspect. Sometimes a block of iron would exhibit a large permanent moment and at other times would have zero moment. The answer lies in the domain concept.

(1) The Existence of Domains. A region of a ferromagnetic sample in which the magnetization is essentially uniform is called a domain. Domains are formed to reduce the total energy of a sample.

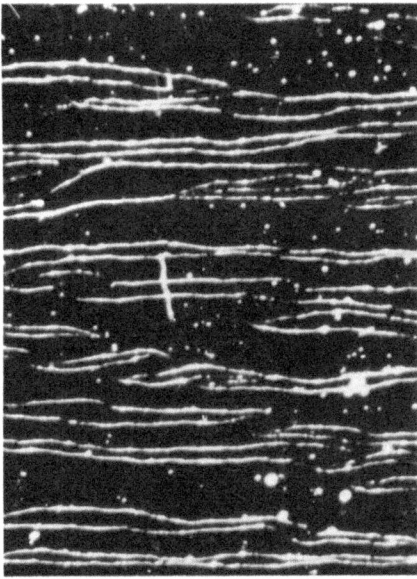

FIGURE 5. Domain walls in a Ni-Fe film observed using the Bitter technique (Mag. = 600×).

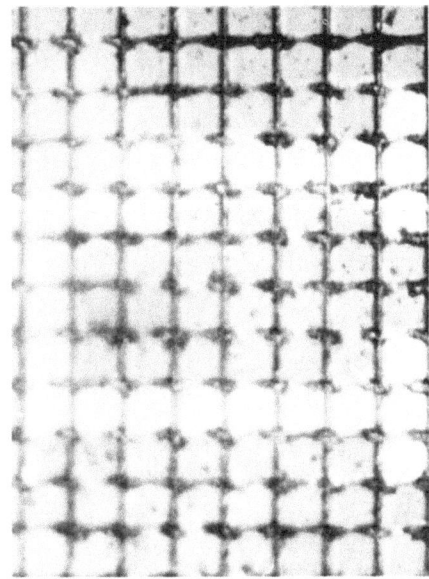

FIGURE 6. Domain in a GdIG wafer observed using the Faraday effect. The wafer is scribed in 30 micron squares. (After Goldberg)

They can be observed directly in a number of ways but one classic method is the Bitter powder method which outlines the boundaries of the domains. In the region between domains, the magnetization configuration leads to stray fields which will attract a magnetic powder. To facilitate the movement of the powder, it is usually suspended in a liquid. A typical picture is shown in Figure 5.

Domains may also be observed by the Kerr and Faraday effects. These utilize the rotation of the plane of polarization of light on reflection from and transmission through magnetic materials. An example of the Faraday effect is shown in Figure 6. For samples thin enough for penetration by electron (< 1200Å for 100 kV electrons), Lorentz Microscopy may be used. Here the internal field bends the electron beam in accordance with the magnetization direction. This technique has become the most power-ful tool for investigations of domains in films. However, at the moment the interpretation is a subject of considerable controversy. It has only recently been recognized that geometric optics cannot in general be used to interpret the results. An example is shown in Figure 7.

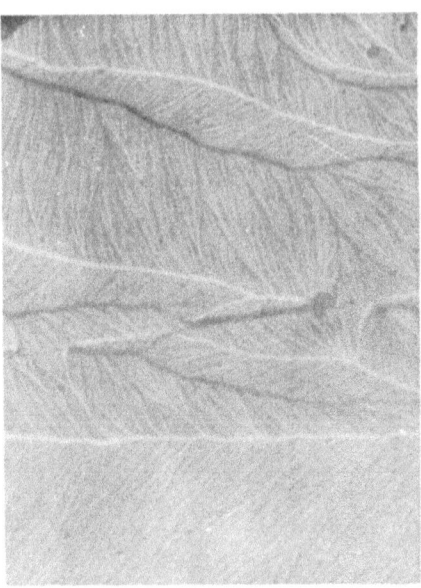

FIGURE 7. Domains and Domain Walls in a Ni-Fe film observed using Lorentz Microscopy. The situation is similar to that shown in Fig. 5. The bright and dark lines are domain walls in which the magnetization on either sides result in a converging and diverging beam respectively. The fine structure is magnetization ripple. (Part 4)

(2) D o m a i n   W a l l s . I mentioned domain walls above. They are the boundary region between domains. Investigations on thin films have tremendously increased our understanding of the complexity of domain wall. The two simplest wall structures are the Bloch wall and the Neel wall. They are shown in Figure 8.

How wide is a wall? Consider a row of spins in a uniformly magnetized sample. If we try to reverse half of them, abruptly, we see that at the interface we pay a considerable amount of exchange energy.

Since $E_x = -2JS^2 \cos\varphi$, for a pair of spins with $\varphi = 180°$, $E_x = +2JS^2$. However, if φ is small, the energy can be written as $E_x = -2JS^2(1 - \frac{\varphi^2}{2})$ or considering only the variable part $E_x = JS^2 \varphi^2$. Thus, if we spread the roation of 180° over N + 1 spins then

$$E_x = NJS^2 \frac{\pi^2}{N^2} = JS^2\pi^2/N \qquad (8)$$

Therefore, the total exchange energy decreases as the wall gets wider.

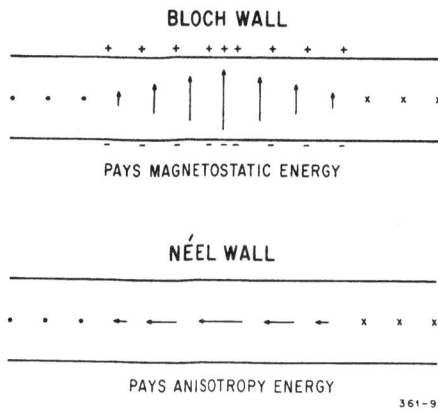

FIGURE 8. Schematic representation of the spin configuration in a Bloch wall (Fig. 8a) and in a Neel wall (Fig. 8b).

As the wall gets wider, more spins are forced to lie at angles to the easy directions of anisotropy. Thus we increase the anisotropy energy as the wall gets wider. This limits the width of the wall.

Let's calculate the energy per unit area and the width of a 180° Bloch wall. Consider a cubic lattice with constant a. Then there will be $1/a^2$ lines of spin per unit area so that from (8) the exchange energy per unit area σ_x is

$$\sigma_x = \pi^2 \, JS^2/Na^2 \tag{9}$$

This anisotropy energy per unit area σ_a is $\sim KNa$. This is very crude but suffices for our purpose. The total energy per unit area σ_w is

$$\sigma_w = \sigma_x + \sigma_a = \pi^2 \, JS^2/Na^2 + KNa \tag{10}$$

To find the equilibrium value of N_1

$$\frac{2\sigma_w}{2N} = 0 \quad \frac{-\pi^2 JS^2}{a^2 N^2} + Ka$$

or

$$N = \left(\frac{\pi^2 JS^2}{Ka^3}\right)^{\frac{1}{2}} \alpha \sqrt{\frac{A}{K}} \tag{11}$$

For Fe, $N \sim 300$ lattice constants.
We can also calculate σ_w

$$\sigma_w = 2\pi \left(\frac{KJS^2}{a}\right)^{\frac{1}{2}} \alpha \sqrt{AK} \tag{12}$$

For Fe, $\sigma_w \sim 1$ ergs/cm^2.

Thus we see that domain walls have quite a finite size and, furthermore, require the expenditure of energy. Both of these properties become particularly important as the sample size becomes very small. In fact, in very small particles, domain walls are unfavorable so that the particles behave as single domains. Most permanent magnets are compact of fine particles. This explains why they have very high coercive forces.

(3) C o e r c i v e F o r c e . As we shall see later, the coercive force in a material with no domains should be $\sim 2K/Ms$. In most materials, Hc is much less than this because the magnetization reverses by domain wall motion.

Consider a perfect uniaxial material with two domains. If we apply an infinitesimal external field parallel to the axis, then it is clear that the wall should move across the sample. Since we have assumed a perfect sample, σ_w is a independent of position and the wall moves to increase the material with magnetization favoring the field. Thus H_c would be zero. In fact, the coercive force is never zero. The energy σ_w is a function of position. This is due to the normal imperfection of materials. The value of H_c is proportional to $d\sigma_w/dx$. Variation in K due to composition or stress gradients will make a contribution to H_c. Bumps which cause the wall to elongate as it passes by increase H_c. In the absence of external field, the wall rests in a local minimum. Then the applied field must be large enough to raise the wall over the next maxima.

The coercive force H_c for wall motion is then not a fundamental quantity but one which very directly depends on material preparation, heat treatment, mechanical damage, etc. Its importance in ferrite core technology cannot be underestimated.

Part III

I. FERRITE CORES

A. *Chemistry*

Because of the square loop behavior of many ferromagnetic materials, it was natural that they should be adopted for storage devices. After several attempts using metallic strips and loops, the ferrite core was developed and has stood almost unchallenged for 15 years.

A ferrite [1] with a spinel structure has the general formula Me Fe_2O_4 where Me represents a divalent ion. In simple ferrites, Me is one of the transition elements Mn, Fe, Co, Ni, Cu and Zn. Almost universally commercial ferrites are Mn and another ion. However, to achieve higher curie temperatures, some ferrites without Mn have been discussed. The normal preparation procedure is to start with oxidic raw materials, mix them and prefire them. They are then remilled, granulated, and pressed into the desired form. This form is usually a torroid.

B. *Anisotropy*

While magnetostriction can influence the magnetic properties, the anisotropy in commercial ferrites comes almost entirely from crystalline anisotropy.

The material generally has negative crystalline anisotropy so that the (111) are easy directions. Since the samples are polycrystalline, after saturation the magnetization will try to lie along the nearest (111) direction. This leads to a theoretical remanence of ~ 0.87. For Mn Fe_2O_4, $K_1 = -28 \times 10^3$ ergs/cc.

C. *The Operation of a Core*

The torroid is a closed flux geometry. Thus one can make it very small without the problem of demagnetizing effects. When the torroid is magnetized circumferentially in one direction, we can call that a o n e and in the other direction a z e r o. Fields are provided from equal currents in two orthogonal wires which pass through the center of the core. The currents are such that the core is switched only when both currents are applied. Thus, we can selectively write (switch) one core in a square array. To read, the same currents are applied. An additional sense line is strung through the cores to detect any flux changes. If the coincident currents for writing a zero are applied to a particular core to read, an output means a one, no output a zero. Obviously this is destructive, so cores must be rewritten after reading.

D. Thermal Characteristics

Ferrite cores are ceramics. They have been fired at very high temperatures ~1200°C. Therefore, they are stable at room temperature and do not suffer from any low temperature aging effects. However, since the anisotropy is crystalline and since the Curie temperature in many cores is low, the properties are temperature sensitive. Figure 9 shows the temperature dependence of M_s. Since K_1 goes like a very high power of the magnetization, it would be advantageous to use a ferrite with a high Curie temperature. This is one of the areas where research may improve the performance of cores.

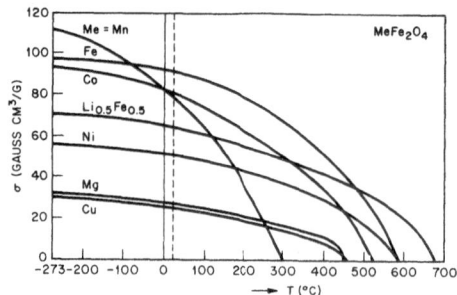

FIGURE 9. The temperature dependence of the magnetization per unit volume for several ferrities (After Smit and Wijn)

E. Dynamic Switching

1. The Reciprocal Switching Curve. In order to operate a core at high speeds, it must be possible to reverse the magnetization rapidly from one sense in the torroid to the other. Therefore, a fundamental measurement is the switching speed τ_s versus applied field. The results are normally plotted as $1/\tau_s$ versus H. A typical set is shown in Figure 10. The three region curve is quite characteristic of switching phenomena in general and has been observed in films as well as cores. Clearly, we can switch the core very fast if we drive it very hard. However, our device is limited to the use of low coincident current. Thus practically we are forced to operate in region I. This is shown expanded in Figure 11. It is now generally believed [2] that switching in region I (and, therefore, in most applications) is due to wall motion so it is worth our while to review the process in detail.

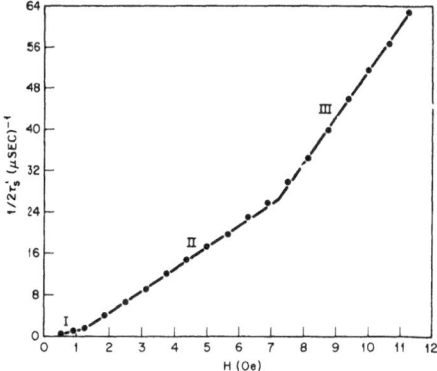

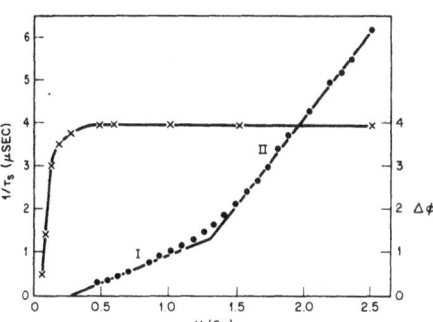

FIGURE 10. Inverse switching time $1/\tau_s$ versus applied field H in a ferrite core (After Gyorgy).

FIGURE 11. Inverse switching time $1/\tau_s$ and switched flux $\Delta\phi$ versus applied field H at low fields (after Gyorgy).

2. Domain Wall Velocity. Consider a uniaxial material with a domain wall parallel to the easy axis. In the absence of an applied field, it rests in an energy minimum. For very small values of the applied field, the wall will now move reversibly. We can write a simple equation of motion for the wall in this case [3].

$$m_w \ddot{Z} + \beta \dot{Z} + \alpha Z = 2M_s H \tag{1}$$

$2M_sH$ is the pressure on wall due to the field H. Since walls in motion have additional energy, we can speak about the mass of the wall and m_w is the mass/area. Since the wall dissipates energy, we need a damping term and β is the viscous damping coefficient. The resorting force is characterized by α which is a material parameter.

When the field H exceeds H_c, the wall motion is irreversible. Then $m_w \ddot{Z}$ can be ignored. The term αZ is replaced by subtracting an effective field H_c from H. Then the equation of motion becomes

$$V = \frac{2M_s}{\beta} (H - H_c) \tag{2}$$

where $2M_s/\beta = R$ is called the mobility.

If the number of walls during a switching process is constant, then we expect from Eq. (2) that $1/\tau$ will increase linearly with H as observed. How then can we increase wall motion switching speeds? We can a) decrease the damping constant, b) increase the number of walls.

3. The Viscous Damping Constant β. In thick metal samples, the flux change during wall motion is sufficient to cause significant eddy currents. The field from these currents opposes the applied field and so slows down the wall. Early attempts to use metal cores failed because by the time the cores were thin enough to reduce the eddy currents, they

were impractically fragile. The ferrite as we have mentioned is an insulator so that eddy currents are not significant.

Measurement on ferrites however showed that β was not zero. In fact, in Fe_3O_4 in a sample in which the estimated value of the eddy current damping of $\beta e \simeq 0.078$, the total measured β was 0.484. This has been attributed to an intrinsic damping mechanism which is not well understood.

Consider in detail how a wall moves. When the field is turned on, the magnetization at every point in the walls begin to precess about the field with a frequency $\omega = \gamma H$. Here γ = gyromagnetic ratio = (ge/2mc) = $2.8 \times 10^7 sec^{-1} oe^{-1}$. This leads to a demagnetizing field perpendicular to the wall. The magnetizations then precess about the net field $(\bar{H} + \bar{H}_D)$. The result is an apparent motion of the "wall".

The magnetization acts like a top with equations of motion like

$$\frac{d\bar{M}}{dt} = -\gamma (\bar{M} \times \bar{H}) \tag{3}$$

where $M \times H$ is the applied torque, $\bar{M}/\gamma = \bar{J}$ = angualr momentum. The solution of Eq. (3) says that the angle between $\bar{M}$ and $\bar{H}$ remains constant but $\bar{M}$ precesses about $\bar{H}$ with a frequency $\omega = \gamma H$. Now it is clear that in a block of iron, when a large field is applied, the magnetization will align itself parallel to H (the lower energy state). What happened to the potential energy? Why didn't M simply precess about $\bar{H}$. In our iron block, we can say most of the loss is in eddy current. But in our ferrite?

Landau and Lifslutz proposed that a term be added to Eq. (3):

$$\frac{dM}{dt} = -\gamma (M \times H) - \frac{\lambda_0}{M^2} (M \times M \times H) \tag{4}$$

λ_0 is a phenomenological damping constant called the relaxation frequency which cannot be derived from basic principles. This term expresses the fact that damping gives rise to a motion perpendicular to $\bar{M}$ and also $\bar{M} \times \bar{H}$.

In the formulation, we assume that the precession term is large compared with the damping and that damping acts only on the precession. Strictly speaking, it should act on the total motion, $d\bar{M}/dt$. Gilbert proposed such an equation which is equivalent to Eq. (4) when the damping is small. His damping term is $\frac{\lambda'}{M}(M \times \frac{\partial \bar{M}}{\partial t})$. If λ_0 is set equal to $\gamma \lambda' M$, and $\lambda'^2 \ll 1$, the two terms are identical. For Fe_3O_4, $\lambda_0 = 3.5 \times 10^8 sec^{-1}$ which gives $\lambda' \sim 10^{-2}$. The rate of dissipation will be $\bar{H} \cdot \frac{d\bar{M}}{dt}$ or $H \cdot \frac{d\bar{M}}{dt} = -\gamma \left[\bar{H} \cdot (\bar{M} \times \bar{H}) \right] - \frac{\lambda_0}{M^2} \left[H \cdot (M \times M \times H) \right]$. Rate of Dissipation $= -\frac{\lambda}{M^2} \left[(\bar{H} \times \bar{M}) \cdot (\bar{M} \times \bar{H}) \right] = +\frac{\lambda}{M^2} (\bar{M} \times \bar{H})^2 \tag{5}$

As a wall moves through the sample a distance ΔX it must dissipate the energy gained = 2MH area ΔX. Therefore, the rate of dissipation per unit area of wall = 2MHV. If we calculate Eq. (5) for a 180° Bloch wall moving with a velocity V, we find that

$$2MHV = \frac{\lambda_0}{\gamma_2} \left(\frac{K'}{A} \right) \frac{1}{2} V^2 \qquad (6)$$

$$V = \frac{2M_sH}{\frac{\lambda_0}{\gamma_2} \left(\frac{K'}{A} \right) \frac{1}{2}}$$

Comparing this with our result for wall velocity (Eq. (2)), we see that

$$\beta = \frac{\lambda_0}{\gamma_2} \left(\frac{K'}{A} \right) \frac{1}{2} \qquad (7)$$

Note that β is inversely proportional to wall width. It is also proportional to λ_0. What is "λ_0"?

This is not clear. It does appear to depend on the motion of the magnetization. One way for example that energy can be dissipated in the lattice is through spin waves. This converts the magnetostatic energy to demagnetizing, crystallizing anisotropy and exchange in the spin wave wake.

Because we are not clear on the origin of β, it is not possible to adjust the composition of the ferrites in any intelligent way. However, the empirical evidence is not encouraging.

TABLE II. The Switching Coefficient S_w for a number of Ferrites (after Gyorgy)

CHEMICAL COMPOSITION	S_w (oe μsec)
$Zn_{0.20}Mn_{0.10}Mg_{0.70}Fe_{1.90}O_4$	0.20 } a 0.42
$Mn_{0.46}Mg_{0.71}Fe_{1.76}O_4$	0.25–0.35
$Ca_{0.06}Mn_{0.58}Mg_{0.67}Fe_{1.84}O_4$	0.22–0.35 } 0.47–0.52 } a
$Zn_{0.39}Mn_{0.47}Mg_{0.34}Fe_{1.62}O_4$	0.22
$Zn_{0.02}Mn_{0.46}Mg_{0.79}Fe_{1.62}O_4$	0.21 } 0.30 } a
$Zn_{0.08}Mn_{0.64}Mg_{0.28}Fe_{1.74}O_4$	0.23
$Zn_{0.18}Mn_{0.57}Mg_{0.16}Fe_{1.79}O_4$	0.33
$Mn_{0.52}Mg_{0.67}Fe_{1.80}O_4$	0.30–0.70
$Zn_{0.69}Ni_{0.29}Fe_{1.98}O_4$ b	0.25 } 0.51 } a
$Y_3Fe_5O_{12}$	0.30

In Table II, we list the switching coefficient $S_w = \tau H$ for several ferrites. It can be seen that they are all between 0.2- 0.5 oe μ sec.

We have overlooked one difficulty in Eq. (4). If we consider the Gilbert form of the damping which is a better description, then for no damping, $\bar{M}$ precesses at constant angle so that $T = \infty$. For infinite damping, $\bar{M}$ doesn't move, and again $T = \infty$. Gyorgy argued that there must be an optimum value of λ'. He showed that the minimum value of S_w should be ~0.2 oe- μ sec.

Thus, there is a theoretical argument which says that we have achieved about the minimum value of S_w.

4. The Number of Domain Walls. We can decrease the effective value of S_w, S_w' by increasing the number of walls [4]. If S_w is for one wall then

$$S_w' = \frac{S_w}{N} \tag{8}$$

where N is the number of walls.

Domain walls must be nucleated. It has been argued that grain boundaries are the chief source of nucleation sites in ferrites because of the demagnetizing fields at the boundary. In Table II, the variation of S_w from 0.2 to 0.7 oe - μsec observed for a Mn Mg ferrite was obtained by increasing the grain size from 2 to 1,500 microns. This does not suggest that the number of nucleation sites is at all linear with the number of boundaries. Of course pores, inclusions, etc., will also serve as nucleation sites and considerable work has been devoted to empirical studies trying to increase the number of sites.

At the present time, the general trend is to decrease the grain size. I feel that this will be of limited value. As grain size decreases, the magnetization will be more strongly coupled across the boundaries and the boundaries will no longer act as nucleation sites. Clearly, in thin films, domains do not nucleate at all grain boundaries, if at any of them. In fact, domain walls cross the boundary without disturbance. I am therefore not convinced that a significant improvement can be obtained in the value of S_w for ferrites.

5. Core Size. A practical improvement in core performance can be achieved by reducing the core dimensions. If the coercive force is kept constant, then a lower power memory can be attained because the current requirement to achieve the drive field is reduced. If the current is kept constant, H_c can be increased and if the ratio of H/H_c is kept constant $(H - H_c)$ will be larger so that S_w will be smaller.

Cores are now being discussed with O.D. of 12 mils and I.D. of 7 mils. This is the smallest one can get and still fit the wires through the center.

F. *Conclusion*

It appears that the technological limit of cores in the coincident currents mode has been achieved. Fundamental and practical limits have been reached.

Part IV

I. POLYCRYSTALLINE FERROMAGNETIC FILMS

A. *Introduction*

A polycrystalline ferromagnetic film [1, 2] combines interesting technological characteristics with fascinating magnetic properties in an extraordinarily convenient geometry. The result has been more than 2000 publications in 10 years. This effort is now being rewarded in a practical sense. Thin film memories are a reality. Even without this, we have learned a great deal about the magnetic behavior of materials of all kinds both fundamentally and technologically from studying films. What makes them interesting? What are the problems?

B. *General Characteristics*

It was thought for a while that below a critical thickness, a planar magnetic sample would be a single domain on a macroscopic scale. Then the magnetization would switch by coherent rotation. Since all the spins would move together, it would be much faster than wall motion switching where the spins go one at a time.

This hope has been partially realized. It is possible to make single domain films as large as one wants. Unfortunately, it is also possible to put domain walls in the same film. In films less than 300Å they are pure Neel Walls. Above 1200Å, they are Bloch walls. In between they are combination walls. The state of a film is sensitive to history. If one saturates it in a particular direction, it will be essentially a single domain. One can demagnetize it, however, and leave multitudes of domain walls behind. We will return to this difficulty in a few moments.

In addition to showing pseudo-single domain characteristics, another general property of polycrystalline magnetic films is a planar uniaxial anisotropy E_u of the form

$$E_u = K_u \sin^2\theta \tag{1}$$

The origin of K_u is still being investigated. Several mechanisms have been proposed, and I will discuss them in Section 5. The easy axis is determined by the direction of $\overline{M}_s$ during preparation.

We will show below that $H_c \sim \dfrac{2K_u}{M_s}$. This can be achieved in the Ni-Fe alloys near 80% Ni. In this composition range, K_1 is small and the magnetization is zero.

All the work on film memory elements has been directed towards the 80-20 Ni-Fe composition, with the addition in some instances of small quantities of Co and various other metals. The films have been prepared by evaporation, sputtering and electroplating. The properties are extremely sensitive to preparation conditions and much of the early work was involved in achieving uniformity.

Normal conditions for evaporation would be T_s = 250°C, 10^{-5} - 10^{-6} mm of Hg.

The uniaxial anisotropy implies that there exist two equally stable directions for $\overline{M}_s$. This, coupled with the potentially fast switching, made films extremely attractive as memory elements. They could be fabricated in arrays by evaporation, sputtering or plating. Strip lines could be put down close to the element to carry the switching currents. The same strip lines could be used as single turn pickup coils to detect flux changes. Since the films were mirror-like, one could even imagine optical read-out. The possibilities boggled the mind ten years ago.

Well, let's see what went wrong.

C. *Quasi-Static Rotational Switching*

Some years ago, magnetization reversal was always associated with domain wall motion. For materials with very high coercive forces, it was assumed that huge stresses existed and these were used in "hand waving" arguments to explain the data. In 1948, Stoner and Wohlfarth, in what must be considered the classic paper in magnetics, pointed out that these arguments had been extended to stresses in excess of the yield point of most materials. They suggested that the reversal in some cases (particularly for very small particles) was occurring by coherent rotation. In that model, the coercive force is determined not by the inhomogeneities but by the anisotropy. Thus in iron for example, they showed $H_c \sim$

$$\frac{2 \times 4 \times 10^5}{1700} \sim 400 \text{ oe if the anisotropy comes from crystalline effects. It}$$

could in fact be much larger if K came from shape effects.

Developments in the permanent magnet industry since then, provide excellent evidence for the essential correctness of this picture. Its influence on thin film technology has been a dominating one.

The coherent rotation model assumes that the magnetization is uniform at every point in the sample. That is, we can represent the moment of the sample by a single vector $\overline{M}_s$ equal to the saturation magnetization. Furthermore, it assumes that the direction of $\overline{M}_s$ will always be such as to make the state of the sample a local energy minimum. The system can then be treated quantitatively.

Consider a film in the x-y plane with uniaxial anisotropy. Because of the dimensions, $\overline{M}_s$ will be in the plane during any quasistatic process. (This is not true in the dynamic case as we shall see.) Let $\overline{M}_s$ make an angle θ with the x axis.

212

The energy density is

$$E = K_u \sin^2\theta - H_x M_s \cos\theta - H_y M_s \cos(90 - \theta)$$

(2)

$$E = K_u \sin^2\theta - H_x M_s \cos\theta - H_y M_s \sin\theta$$

The critical equilibrium conditions are

$$\frac{\partial E}{\partial \theta} = 0 ; \quad \frac{\partial^2 E}{\partial \theta^2} = 0 ;$$

(3)

$$\frac{\partial E}{\partial \theta} = 0 = 2K_u \sin\theta \cos\theta + H_x M_s \sin\theta + H_y M_s \cos\theta$$

$$\frac{\partial^2 E}{\partial \theta^2} = 0 = 2K_u \cos 2\theta + H_x M_s \cos\theta + H_y M_s \sin\theta$$

If we divide by M_s and let $\dfrac{2K_u}{M_s} = H_K$

then

$$H_x^c = -H_K \cos^3\theta$$

(4)

$$H_y^c = H_K \sin^3\theta$$

We recognize this as the equation of an a s t r o i d (Figure 12). At these values of H_x and H_y, the number of stable positions available to the magnetization changes from two to one. This is not obvious from Eq. (4). In fact, any value of θ such that $\dfrac{\partial E}{\partial \theta} = 0$ and $\dfrac{\partial^2 E}{\partial \theta^2} > 0$, is stable. This can be shown by numerically solving the parametric equation which one obtains from Eq. (2). However, the physics of the rotation model are more clearly seen from the astroid itself. It can be shown rigorously that the stable positions of M_s can be obtained by drawing a line from the top of $\bar{H}$ tangent to the astroid. $\bar{M}$, parallel to this line, is stable.

From the astroid, we see that in the hard direction, the hysteresis loop should be linear for $H \leq H_K$ and reversible. In the easy direction, the loop is square with $H_c = H_K$.

This exposes several attractive features of our theoretical film.

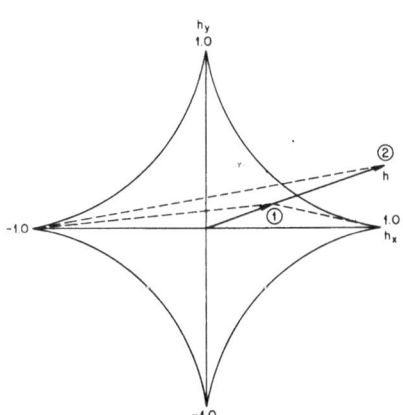

FIGURE 12. The critical switching curve, the astroid, for a uniaxial single domain material.

1) Since the hard direction loop is reversible for $H < H_K$, we can read non-destructively.

2) We can switch by applying a field in the easy direction coincident with a field in the hard direction. This hard direction write field can be the same magnitude as the hard direction read field.

D. Dynamic Switching

The coherent model tells us the "before and after" position of the magnetization but gives no insight into how the magnetization moves.

From our discussion of wall motion, however, we can follow that process easily.

Imagine a film saturated along the easy direction. If we apply a step function field $H > H_K$ in the hard direction eventually M_s will try to precess about H with a frequency $\omega = \gamma H_D$ where γ is the gyromagnetic ratio = 2.8×10^7 sec^{-1} oe^{-1}, and H_D is the demagnetizing field normal to the film plane.

As we saw in domain wall switching, the damping frequency λ will determine the switching time τ. The minimum switching time will occur when the system is critically damped. Although the films are metal, because of their thickness there will be only a small contribution to the damping from eddy current.

Using the Landau-Lifshutz equation, it is possible to calculate the switching process in some detail. Let's avoid that and spend our time elsewhere. Some simple ideas will suffice.

Since
$$\omega = \gamma H_D \tag{5}$$

and since the demagnetizing field $H_D = 4 \pi M_s \sin \alpha \sim 100$ oe, where α is the angle between $\overline{M}_s$ and the film plane, then $\omega \sim 2.8 \times 10^7 \times 10^2 \sim 3 \times 10^9$ sec^{-1}.

Therefore, $\tau \sim 1$ - 10 nanoseconds. Experimental studies have revealed that the oscillation of $\overline{M}$ is nearly critically damped and values of τ approaching 1 nanosecond have been achieved. Thus film performance is not limited at present by switching time.

Now let me make it clear that theory and experiment are not in good agreement when one looks in detail. We still do not understand the value of λ or the field dependence of τ. To understand it, we must use a spin wave treatment. However, films do switch very fast and their limitations lie elsewhere.

E. Coercive Force, Domain Walls, and "Creep"

As soon as the first workers were able to build hysteresis loop tracers, the first major deviations from the coherent model were observed. Although, at low drives the hard direction loop was linear, as M approached M_s, hysteresis was generally observed. By extrapolating the linear portion to M_s, the first values of H_K were obtained. Depending on thickness and composition, they ranged from 1-10 oe.

More surprising was the observation that although the easy loop was square, $H_c < H_K$. It is now clear that the reason for this is that when the reversing field is near the easy axis, reversal occurs by the motion of domain walls. The critical curve must be modified, but this in itself does not seriously interfere with films as storage elements. The write field in the hard direction is now used to snap M_s into the hard direction

214

and then a small easy field is put on to determine the direction of fall of M_s when the hard field is shut off.

The presence of domain walls has a much more serious effect, however, than simply requiring us to modify the read-write operation. It is nearly impossible to prevent the formation of domain walls in any practical memory. If one uses small discrete islands, the walls form at the edges where the demagnetizing fields are enormous. If one uses a continuous sheet, then the bits are naturally bounded by domain walls.

The problem is that the threshold for wall motion in dc fields is substantially lowered if the hard direction field is alternating. Thus as one writes in one location, stray fields can destroy information in nearby locations.

Creep is characterized by the following responses:

1) Creep occurs only (and equally) during the rise and fall of the hard field.

2) Creep is independent of rise time down to a few nanoseconds.

3) Creep is thickness-dependent. Very thin films are least sensitive.

4) Creep occurs more readily in bipolar than in unipolar hard fields.

5) The average wall jump distance per pulse increases rapidly with easy field.

Generally, theoretical attempts to explain creep have tried to find a mechanism which generated a field parallel to the easy direction. This field would add to the applied easy field so that the net field exceeded H_c. At the present time, there is no accepted model although several have been suggested. The problem is minimized by spacing the bits far enough apart to prevent the stray field at one location from causing creep at another.

F. Magnetization Ripple

Numerous studies of magnetization reversal demonstrated that the coherent rotation model was a poor description of a film. Because of the historical origin of the model in small particle work, it was, I suppose, natural to replace the film by a collection of non-interacting regions each of which had slightly different characteristics. From this came the concept of anisotropy dispersion.

In retrospect, I feel that this was extremely unfortunate and has led to a considerable misunderstanding of film behavior. Ideas derived from the picture have been applied in situations which clearly have little relationship to the model but are by now firmly rooted by tradition.

I remind you that we are dealing with polycrystalline films with grains ranging in size from $50 \, \overset{\circ}{A}$ to $10,000 \, \overset{\circ}{A}$. Each grain has a particular crystallographic orientation with associated easy axis. This is addition to the uniaxial anisotropy. Can the magnetization ignore these deviations? The answer can be seen very dramatically in a Lorentz Micrograph of a film (Figure 13). The "ripples" are normal to the average direction of magnetization and arise because the local magnetization direction varies around the average direction.

Because there is a much stronger transverse magnetostatic coupling than longitudinal the ripple wavelength normal to the magnetization is much longer than the wavelength parallel to the magnetization. In the longitudinal direction, exchange coupling dominates but is much weaker than the transverse magnetostatic coupling.

FIGURE 13. A Lorentz micrograph showing local variations ("ripples") in the average magnetization direction of a Ni-Fe film.

To properly treat the problem, one must carry out a tedious minimization of all the appropriate energy terms, assuming a stochastic variation in the orientation of the crystallites. This work has occupied several theoretical groups for a number of years. For films in which the local perturbation is small, there is general agreement on the results. However, as the applied field approaches a critical point, or if the local perturbations are large, the ripple becomes very large and the situation is not very good.

It has been very difficult to verify the predictions of ripple theory experimentally. The agreement is qualitative at best. Lorentz microscopy has failed to provide a clear-cut result.

We can conclude that considerably more work must be done before we can understand the switching characteristics of films. This understanding will become even more important when memories which operate near the switching time limit of the film are built.

G. The Anisotropy

The last of the three most serious problems in film physics is the induced anisotropy. We have pointed out that if one evaporates a ferromagnetic metal in an applied field, then the resultant film has an anisotropy like $K_u \sin^2\theta$. The same is true for electroplating and also for sputtering and, as far as one can tell, for any method you use.

K_u versus composition on NiFe shows a minimum near 82% Ni. This strongly suggests some kind of stress mechanism. However, even at $\lambda = 0$, $K_u \neq 0$ and it is generally believed that iron-pair ordering contributes to K_u. At the present time, the situation is not clear and both mechanisms probably contribute.

First let me point out the importance of the problem. It is true that K_u in NiFe has a value appropriate for a film storage device. However, a serious question of the temperature stability of K_u has been raised. In particular, will films change their properties at temperatures of $\sim 75°C$ after several years of operation?

H. Constraint Release

A free standing magnetostrictive film will deform to lower its energy. If we take such a deformed film and glue it to a rigid body, and if we try to change the magnetization direction, the film will exhibit an anisotropy. The bonding will prevent the film from deforming to adjust to the new magnetization position. The film will, therefore, try to keep the magnetization where it was and a uniaxial anisotropy will result.

It is clear how the model applies to an evaporated or sputtered film. In fact, it appears to make a substantial contribution in Fe and Ni films and in NiFe films below 70% Ni. Here the magnetostriction is large. However, we are interested in films with zero magnetostriction so that the anisotropy must have its origin elsewhere.

I. Pair-Ordering

Consider a NiFe alloy, e.g., 75-25. In the absence of magnetic effects, e.g., when $T > T_c$, one would expect to have as many Fe-Fe, Ni-Ni and Fe-Ni atom pairs in one direction as in any other. Now, if we lower the temperature below T_c, we can assign an anisotropy to each pair. That is, each pair would rather have the magnetization along it than perpendicular to it. If, further, the pairs do not have the same anisotropy, then we would expect that the system would try to adjust the pair distribution to lower the energy. This would result in a macroscopic uniaxial anisotropy. The magnitude of the anisotropy will be determined by the interdiffusion of pairs and will obviously be temperature sensitive and annealable.

The problem with the model is that we don't know the pair energy. Theoretical estimates are $\sim 10^{-15} - 10^{-16}$ ergs, and this is the order of magnitude required to explain bulk data on Ni-Fe. The agreement in detail is only fair. It is able to give a reasonable description of the compositional dependence but the predicted magnitude as a function of anneal direction in a single crystal is too small.

What about in films? The first problem is that in bulk the relaxation time at 350°C is ~ 100 seconds. Furthermore, many investigators have examined the annealing behavior and found very fast, very low temperature effects. Substantial changes can occur in 15 min at 200°C. Most workers are convinced that in films we are faced with several different kinds of defect ordering. They may be vacancies, impurities, etc. The usual experiments have failed. Measurement as a function of time and temperature do not fit any unique set of processes.

Fortunately, it has been found that if films are annealed at elevated temperatures, they can be stabilized. This is now a routine procedure for electroplated films.

J. Conclusion

Films can be used to make fast memories. We understand an enormous amount about them but we still have several major problems. Fortunately, they have fundamental as well as technological interest.

To solve the creep problem, we must influence the wall structure.

To understand switching, we must understand ripple.

To understand the anisotropy, we must investigate the defect structure.

Part V

MEMORIES OF THE FUTURE

In this final part, I intent to discuss two memory schemes which are still very much laboratory devices. The first, the magneto-optic memory, is the most advanced. The second, the photo-chromic memory, is still a dream. Both rely on our present ability to form a small intense spot of light with a laser. Both use a light beam to read and write information. Thus we might have hope to make a large capacity, high speed memory.

I. MAGNETO-OPTIC MEMORY

If a beam of linearly polarized light passes through a material, then the plane of polarization will be rotated by an amount $\theta = \beta L M_z$. Here β is the Faraday constant, L is the length of the sample and M_z is the component of the magnetization along the light beam.

This effect (and the Kerr effect) have often been suggested as a means of reading magnetic material. However, if one must still write by conventional means, the device has no particular advantage for main frame storage. If, however, an effect could be found which caused the coercive force to be temperature sensitive, then one could selectively write by lowering the coercive force locally with the heat from the light beam in an applied field.

A material which has this property is GdIG [1]. The temperature dependence of the magnetization is shown in Figure 14. It has a compensation temperature at about 15°C. The reason for this is that the two sub-lattices Gd and Fe are anti-ferromagnetically coupled and their magnetizations have different temperature dependencies. Now since we expect $H_c \sim K/M$ then at T_c, $H_c \longrightarrow \infty$. More realistically, we expect that H_c will have a peak at T_c. This has been observed. For a single crystal, $H_c > 100$ oe at T_c but drops to ~ 20 oe ± 5°C away.

In order to get large rotations, we would like the easy axis in the beam direction. This is the (111) in a single crystal. In order to get large rotations, we would like thick samples but to reduce absorption we want thin samples, so we must choose an optimum thickness. The absorption coefficient α is a function of wavelength. The ratio of β/α for GdIG is shown in Figure 15. One sees that the optimum is > 7000Å while 5600Å is a second choice. Unfortunately, one wants to use a laser so we are limited to discrete wavelengths.

A possible choice is the 5680Å line of an argon laser. However, we would also like the sample thin to reduce the thermal requirements. Since the optimum thickness for signal to noise is wavelength dependent, it is possible that the best overall choice is the 5145Å line of the same laser. Here the optimum thickness is 10 microns.

How does the memory work (Figure 16)? We write by applying a field over the whole sample and hitting a spot with a laser beam. This lowers H_c at the spot, the region reverses. We shut off the laser. To read, we simply detect the direction of rotation of the plane of polarization.

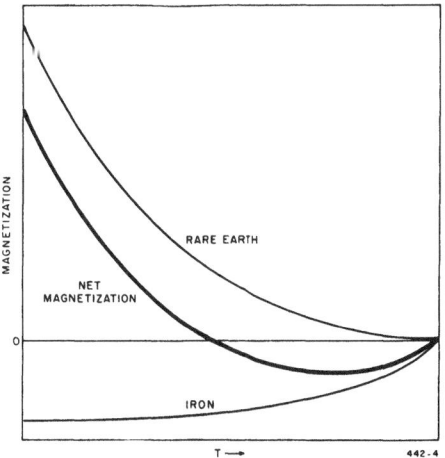

FIGURE 14. Temperature dependence of the net magnetization and of the sub-lattice magnetization in GdIG.

As in any new device, there are problems. Here are the serious ones:

A. Minimum Spot Size. Because of the low value of M near the compensation point, only a little energy is gained if the magnetization switches to be parallel to the applied field. Moreover, a domain wall must form which costs some energy. The final domain size is determined by the balance between these energies. We can calculate the minimum nucleation radius r as follows:

$$2\pi r^2 \, LMH_c = 2\pi rL\sigma_w$$

or

$$r = \sigma_w / M H_c$$

Substituting appropriate values, r ~ 40 microns. This is too large for a practical device. We would like r to be 1 micron.

One method which has been used to reduce the domain size is to scribe the wafers (Figure 6). However, it is doubtful that very small spots can be achieved with this technique. Recently it has been reported that polycrystalline films have spot sizes of the order of crystallite size. This is a major advance and could be the solution to the problem.

B. Samples. Polishing single crystals is expensive and impractical. The device will require thin films probably prepared by sputtering.

C. Heat Limit. This is estimated to limit the speed to ~ 1 microsecond. If the required temperature change could be reduced, the speed would increase. Thus we would like to increase the sensitivity of H_c to temperature.

D. Digital Light Deflector. This is the major obstacle to a practical magneto-optic memory. Present electro-optic materials require high power so that the deflector is too expensive. New materials like $Ba_x Sr_{1-x} Nb_2O_6$ where $0.25 \leqslant x < 0.75$ hold considerable promise, and research in this area is intense.

Some schemes using electron beam writing and optical reading have been suggested to eliminate the light deflector.

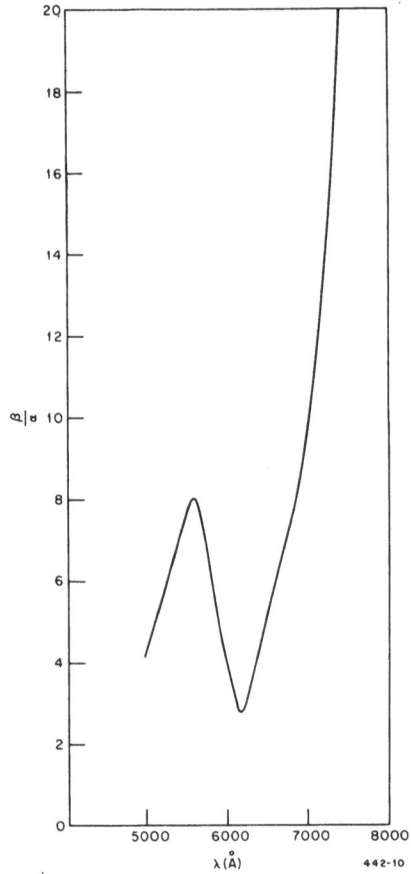

FIGURE 15. The ratio of rotation β to absorption α in GdIG versus wavelength (After Goldberg).

II. PHOTO-CHROMIC MEMORY

The suggestion has been made often to use the absorption of color centers in crystal to store information. Most schemes use the density of color centers but these are unfortunately affected by the "read" operations.

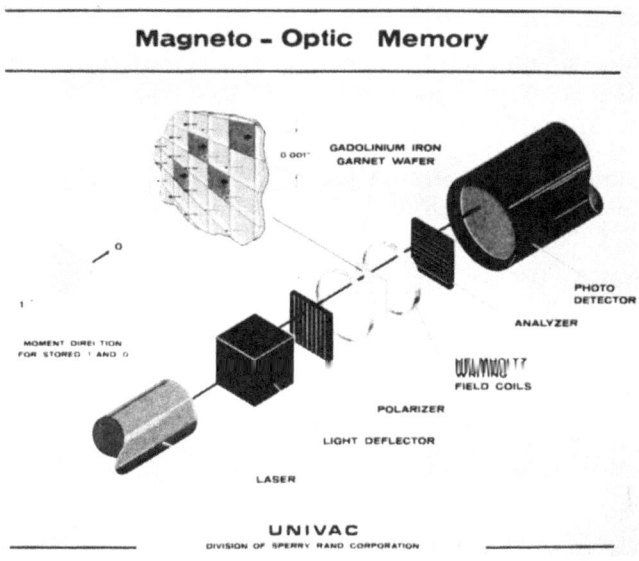

FIGURE 16. A schematic diagram of a magneto-optic memory (After Goldberg).

Recently, a scheme to use the orientation of the color center, which relates the stored information to the dichoric nature of the crystal, has been proposed [4]. It would use an alkali halide crystal, e.g., KCl, which is cheap and easy to produce.

An F color center is an electron trapped at a negative ion vacancy. An M_A center consists of a pair of nearest neighbor F centers lying next to a Na^+ impurity. The Na^+ is more or less fixed in the lattice and the Ma center can re-orient around this impurity.

If a KCl crystal is exposed to polarized light at say 77°K, then almost all the M_s center will align along a single $< 110 >$ lattice direction. Now to write we shine light of $\lambda \sim 5000\text{Å}$, polarized to $< 110 >$ and we will re-orient Ma center to a new $<110>$ direction. We can read with light $\lambda = 8200\text{Å}$ polarized perpendicular to original $<110>$. The dichoric property will allow us to distinguish ones from zeros.

Is this a good idea? I don't really know. We will have to look at it more carefully. But it does something none of our other schemes do. It uses the defect structure. We use the material on an atomic level. That is very attractive to me. It would be an interesting place to start some research.

221

III. CONCLUSION

We have taken a quick tour through the material problems associated with information storage. This has included acoustics, magnetics, semi-conductors, optics, and in the limit photochromics. There is room for research in each of these areas. In each case, however, the researcher must be willing to accept that, like people, materials are not perfect and it is their imperfection, again like people, which makes them interesting.

REFERENCES

Part 1

1. J. RAJCHMAN, Scient. Am. **217**, 18 (1967).
2. A. MILLER, Electronics, p. 171, (February 20, 1967).
3. A. R. SASS, W. C. STEWART and L. S. COSENTINO, IEEE Trans. Magn., 3, 260 (1967).
4. L. C. HOBBS, IEEE Trans. Elect. Computers, **15**, 534 (1969).

Part 2

1. S. CHIKAZUMI, Physics of Magnetism, Wiley (1964).
2. G. T. RADO and H. SUHL, Magnetism, Vol. III, Academic Press (1963).
3. C. KITTEL, Solid State Physics (Seitz and Turnbull Series) Vol. 3, p. 437.

Part 3

1. J. SMIT and H. P. J. WIJN, Ferrites, Wiley (1959).
2. E. M. GYORGY, Magnetism (Rado and Suhl) Vol. III, p. 525, Academic Press (1963).
3. S. CHIKAZUMI, Magnetism (Rado and Suhl) Vol. III, Academic Press (1963).
4. H. P. PELOSCHEK, Prog. Dielect., **5**, 37, (1963).

Part 4

1. S. MIDDELHOEK, Ferromagnetic Domains in Thin Ni-Fe Films. Thesis, University of Rotterdam (1961).
2. M. PRUTTON, Ferromagnetic Thin Films. Buttersworth.

Part 5

1. J. T. CHANG, J. F. DILLON and U. F. GIANOLA, J. Appl. Phys., **36**, 1110 (1965).
2. N. GOLDBERG, IEEE Trans. Magn., 3, 605 (1967).
3. K. K. CHOW, W. B. LEONARD and R. L. COMSTOCK, IEEE Trans. Magn., **4**, 416 (1968).
4. I. SCHNEIDER, Appl. Optics (December 1967).

9/Semiconductor Photovoltaic Effect and Devices

MORTON B. PRINCE

I. INTRODUCTION

The photovoltaic effect is the generation of a voltage due to optical
excitations when a semiconductor is illuminated at the electrodes or at
internal barriers or p-n junctions. The specific effect that will be the
subject of these lectures is the voltage developed when a semiconductor
body containing a p-n junction is exposed to optical illumination. Two
possibilities exist and both will be considered; one, when one ohmic contact
is on the p-region and the second ohmic contact is on the n-region of the
body; and two, when both contacts are on the same type of material. These
are known as the transverse photovoltaic effect and the lateral photovoltaic
effect, respectively. These are designated in Figure 1. The transverse
photovoltaic effect is commonly used in solar energy converters,
photodiodes, radiation detectors, special types of null detectors and
tracking detectors, etc. The lateral photovoltaic effect is used at present
only in what is commercially known as the radiation tracking transducer.
For most applications, the devices are operated in the steady state
condition with the radiation continuously falling on the devices (or with the
radiation intensity changing slowly compared to the time constants of the
device). Thus most of the discussion will be devoted to the steady state
solutions of the fundamental relations. There are some applications that
make use of the transient response of photovoltaic devices and some
special solutions will be considered.

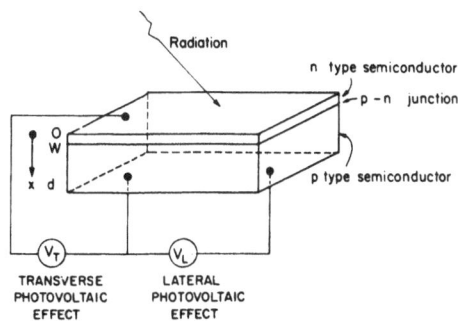

FIGURE 1. Photovoltaic Effects

II. GENERAL ANALYSIS

The general solutions that should be solved in order to obtain the theoretical characteristics of the devices are the three dimensional continuity equations for the excess holes (p) above equilibrium values in the n-type semiconductor region and the excess electrons (n) above equilibrium values in the p-type semiconductor region. The first approximation that will be made is that one dimensional variations only will be considered. Then the equations become the following:

$$\frac{\partial p}{\partial t} = -\frac{p}{\tau_p} + D_p \frac{\partial^2 p}{\partial x^2} - \mu_p \frac{\partial}{\partial x}[(p+p_n) E] + g(x)$$

$$\frac{\partial n}{\partial t} = -\frac{n}{\tau_n} + D_n \frac{\partial^2 n}{\partial x^2} - \mu_n \frac{\partial}{\partial x}[(n+n_p) E] + g(x)$$

where τ_n, τ_p = excess minority carrier lifetimes

D_p, D_n = diffusion constants for holes and electrons

μ_p, μ_n = mobilities for holes and electrons

p_n, n_p = equilibrium values of minority carrier holes and electrons

E = electric field

g(x) = generation term for external exciting sources.

Using the notation of Figure 1, the boundary conditions are:

$$D_p (\partial p/\partial x)_{x=0} = s_p \, p$$

$$D_n (\partial n/\partial x)_{x=0} = s_n \, n$$

$$p(W) = 0$$

$$n(W) = 0$$

where s_p = surface recombination velocity at surface x = 0

s_n = surface recombination velocity at surface x = d

With radiation falling uniformly on the surface x = 0, we can consider the steady state case with $g(x) = \alpha N e^{-\alpha x}$ where $\alpha = \alpha(\lambda)$ is the absorption coefficient which is a function of wavelength and N is the surface density of photons impinging at x = 0.

Also let us consider at this time only the cases where the electric field, E, is non-existent. Then the continuity equations reduce to

$$\frac{d^2 p}{dx^2} - \frac{1}{L_p^2} p + \frac{\alpha N}{D_p} e^{-\alpha x} = 0$$

$$\frac{d^2 n}{dx^2} - \frac{1}{L_n^2} n + \frac{\alpha N}{D_n} e^{-\alpha x} = 0$$

where L_p, L_n = minority carrier diffusion lengths; $L_p = \sqrt{D_p \tau_p}$.

224

These equations can be solved exactly with the above boundary conditions and the solutions can be converted into the following short circuit current relations which are useful for further analysis.

$$Q_n = J_n/N = \alpha L_p \left\{ \frac{e^{-\alpha W}}{\alpha L_p+1} + \frac{e^{(1-\alpha L_p)W/L_p} - \alpha L_p + \frac{S_p L_p}{D_p}\left[e^{(1-\alpha L_p)W/L_p} - 1\right]}{(\alpha^2 L_p{}^2-1)(\cosh W/L_p + \frac{S_p L_p}{D_p}\sinh W/L_p)} \right\}$$

$$Q_p = J_p/N = \alpha L_n \left\{ \frac{e^{-\alpha W}}{\alpha L_n+1} + \frac{e^{-\alpha d}\left[e^{-(1-\alpha L_n)\left(\frac{d-W}{L_n}\right)} - \alpha L_n + \frac{S_n L_n}{D_n}\left(e^{-(1-\alpha L_n)\left(\frac{d-W}{L_n}\right)} - 1\right)\right]}{(\alpha^2 L_n{}^2-1)\left[\cosh\left(\frac{d-W}{L_n}\right) - \frac{S_n L_n}{D_n}\sinh\left(\frac{d-W}{L_n}\right)\right]} \right\}$$

$$Q = Q_n + Q_p$$

where J_n = charge carrier flux reaching the junction from the n-region
J_p = charge carrier flux reaching the junction from the p-region
Q_n = quantum efficiency contribution from n-region
Q_p = quantum efficiency contribution from p-region

It is assumed that the geometry of Figure 1 prevails with the n-type semiconductor region exposed to the radiation. Solutions to these equations have been calculated for various values of the parameters L_n, L_p, D_n, D_p, s_n, s_p, W, and d. The resulting independent parameter, α, is then allowed to range over values that yield some significant quantum efficiency. Plots of these curves will be given. It was found that in most cases, it matters little what values are selected for the surface recombination parameters. In fact, by going to the two extreme limits of s = 0 and s =∞, the plots of the resulting characteristics of Q versus α result in curves that differ only slightly from the more correct calculations except in the cases where the junction is extremely close to the top surface as, for example, the optimized solar cell with a gridded structure (which will be discussed in the next section). Plots of Q versus α are given in Figures 2 and 3 for two cases with the parameters listed on the figures. These parameters are useful ones for the narrow band self filtering detectors that will be discussed later in the lectures. Several facts should be noted from these curves; one, in Figure 2, the range of interest of α is 10 to 3000 cm^{-1} with the peak response occurring for $\alpha = 200$ cm^{-1}; the peak quantum efficiency in this case is about 15%; the upper region gives most of the response and its peak is displaced toward the higher values of α compared to the lower region (this is as expected since the higher absorption coefficient photons are absorbed closer to the top surface of the device). In Figure 3, the total response is only about one half that of Figure 2 since the junction is twice as deep below the top surface; however, of more interest is the fact that a narrower range of α gives the total response (in this case α ranges from 10 to 500 cm^{-1}). The reason for the unusual plots of quantum efficiency versus absorption coefficient is that these plots are semi-universal and all that is needed are the plots of absorption

225

coefficient versus wavelength for various materials to understand their spectral response characteristic. For broad band response as in solar cells, the α should have a relatively slow variation with wavelength compared with a rapid variation in narrow band detectors. Such considerations lead to selection of materials for various types of devices.

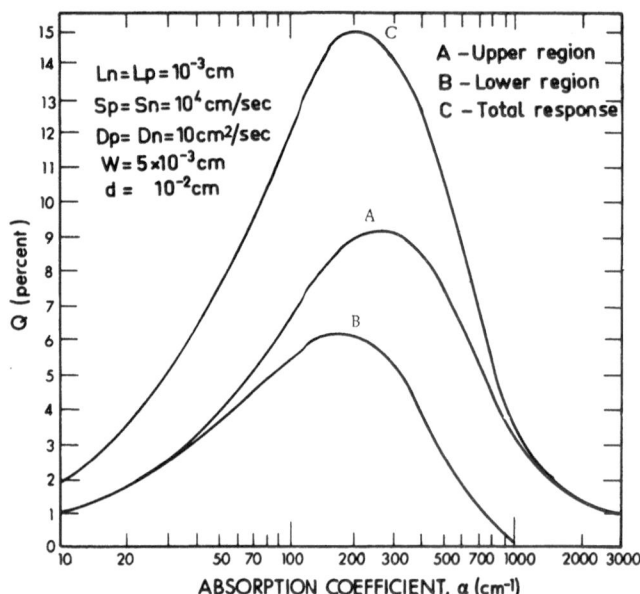

FIGURE 2. Photoresponse Efficiency versus Absorption Coefficient — Case I

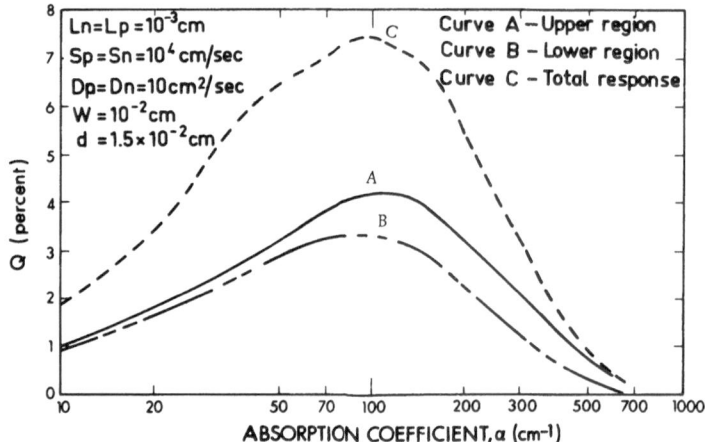

FIGURE 3. Photoresponse Efficiency versus Absorption Coefficient — Case II

III. SOLAR ENERGY CONVERTERS

Using the relations in the last section for the quantum efficiency (number of charged carriers crossing the p-n junction for each photon impinging on the surface of the device), one can optimize the structure of the device and selection of the material (based on absorption coefficient versus wavelength characteristic) for given radiant sources. This optimization is for short circuit operation of the device. For power generation it is essential that the output power of the device be maximized. For the remainder of this section it is assumed that the radiant source is the sun at its zenith and the detector is normal to the solar radiation and located at sea level. This solar cell is a p-n junction which essentially has an electric current generator (I_L) in parallel with the ideal junction, $I_D = I_0 (e^{qV/kT} - 1)$, with a device shunt resistance (R_{sh}) and a device series resistance (R_s) as shown in Figure 4. The equation relating these elements is

$$I = I_0 \left[e^{\frac{q(V+IR_s)}{AkT}} - 1 \right] + \frac{V+IR_s}{R_{sh}} - I_L$$

The parameter A ($A \geq 1$) is introduced since the ideal junction does not exist and empirically this parameter allows the fitting of experimental devices of many materials. R_L is the load resistance across which the power is delivered. When $R_L = 0$, the short circuit condition prevails and $I = I_{sc} \cong -I_L$ when R_s is small (less than 2 ohms). When $R_L = \infty$, the open circuit condition prevails and

$$V = V_{oc} \cong AkT/q \ \ln(I_L/I_0).$$

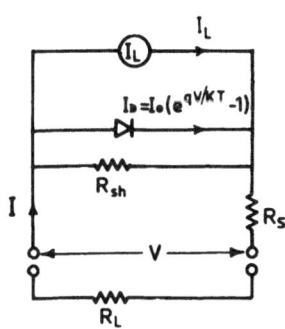

FIGURE 4. Equivalent Circuit of Solar Energy Converter

Referring to Figure 5 which shows plots of current-voltage characteristics the upper curve shows the characteristic in the dark. When exposed to radiation, the characteristic is depressed by a value I_L (light generated current) into the fourth quadrant. Thus the power loss across a load is negative which means that power is generated by the device equal to the area of the shaded rectangle. One maximizes the power output by maximizing the rectangle under the curve. Figure 6 shows plots of the solar cell operation with various values of R_{sh} and R_s and shows how strongly the output power depends on R_s. A series resistance of only 5 ohms will reduce the output power by more than 60% while reducing the shunt resistance to 100 ohms (a highly improbable reduction) reduces the power only a few percent of its maximum value.

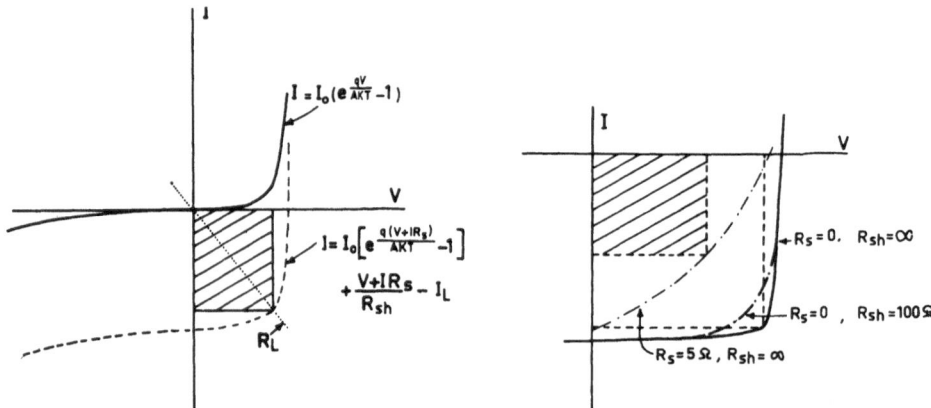

FIGURE 5. Current–Voltage Characteristics of Solar Energy Converter

FIGURE 6. Effect of Series Resistance and Shunt Resistance Variations on Current–Voltage Characteristic

By making approximations for how the various parameters vary with the forbidden energy gap of the semiconductor, it is possible to plot curves of maximum converted power density as a function of the energy gap. Independent of the approximations, the ideal energy gap for 300 °K operation in sunlight is about 1.3 eV with a broad maximum. The peak efficiency is approximately 23% with silicon having a maximum conversion efficiency of 22%. Germanium would yield a maximum conversion efficiency of only 12%. These numbers are based on the power product of open circuit voltage times short circuit current. However at the maximum power point on the curves given in Figures 5 and 6, $V_{max} \sim 0.8\,V_{oc}$ and $I_{max} \sim 0.85\,I_{sc}$; thus even with $R_s = 0$, the maximum power conversion efficiency would be reduced to $\sim 15\%$. Series resistance and other factors reduce this value to the range of $10 - 13\%$ which is the actual production range. The series resistance in the higher efficiency devices has been reduced to approximately one ohm and the thickness of the top layer of the device to approximately one micron by applying a metallic grid structure to the top thin layer which contains most of the series resistance.

Materials other than silicon are not suitable for highly efficient solar energy conversion for one of several reasons:
1) The energy gap is not in the proper range.
2) The absorption coefficient versus wavelength characteristic is too steep.
3) Mixed composition crystals having suitable energy gap and absorption coefficient characteristics are difficult to produce and therefore too expensive for use.

Other aspects of solar cells considered in the lectures are:
1) Temperature variations of the characteristics.
2) Radiation effects and radiation resistant solar cells.
3) Large array considerations.
4) Heterojunctions.
5) Thin film solar cells.

Very briefly the points covered under these five aspects are the following:

1. Temperature Variations

The V_{oc} decreases with increasing temperature primarily because of the rapid increase in I_0 (reverse saturation current) with temperature. The decrease amounts to 2 to 3 millivolts per degree centigrade. The I_{sc} increases slightly with temperature since the forbidden energy gap decreases slightly with increasing temperature and thus some photons of longer wavelength are absorbed and give rise to additional current. The overall effect for silicon devices is a power reduction of 50% upon increasing the temperature by 75°C.

2. Radiation Effects

Operation of silicon solar cells (and cells made from other materials) in space environments has resulted in degradation of the output in time due to radiation damage to the bulk properties (primarily minority carrier lifetimes) by the radiation belts through which the cells pass. It can be shown and has been demonstrated that it is possible to build in the cell an electric field by controlling the distribution of impurities. The resulting impurity gradient electric field can be made as high as 30 volts/cm which gives rise to collection of carriers by drift. So when the lifetime reduction essentially kills the collection of carriers by diffusion, the drift by the built-in electric field prevents a complete collapse of the cell. However in order to produce this concentration gradient of impurities, the cells are made thinner (100 microns) than the standard cells (400 microns) and the initial collection efficiency is lower than that of the standard cells.

3. Large Array Considerations

In making panels of solar cells for significant power generation, it is necessary to arrange the cells in series and parallel groups. In series strings, the string will pass only that current generated by the weakest cell and similarly in parallel groups, the voltage developed is that of the weakest cell or series string of cells. By clever measurement techniques and grouping of individual cells it is possible to obtain about 99% of the individual maximum outputs from an array as compared to about 50% or less for a random arrangement.

4. Heterojunction Cells

Several ideas have been proposed for making more efficient use of the photons in solar radiation by combining materials of different forbidden energy gap. One such idea is to have two or more p-n junctions in materials of different energy gap stacked one upon the other with the highest energy gap material on top. Such a multilayer device can be shown to be not only technologically difficult to make and use but also economically unattractive. A more interesting device would incorporate two different materials into a heterojunction (each material being a different type semiconductor – n-type and p-type) again with the larger energy gap material directly exposed to the incoming radiation. The larger energy gap material would absorb the shorter wavelength photons and the smaller energy gap material would absorb the longer wavelength photons to which the upper layer is transparent giving an overall better photon collection efficiency. Again, however, the technological and economic considerations do not warrant pursuing this development.

5. *Thin Film Solar Cells*

During the last decade considerable effort has gone into the development of solar cells using thin films of amorphous materials. Both CdSe and CdS have been heavily investigated with CdS yielding some very useful results. The device barrier is not a p-n junction which would have too high an energy gap but rather a barrier layer formed by CuS to the CdS. By using the latest technological tricks, cells have been made commercially available with conversion efficiencies up to 6% in areas of 55 cm^2. At present these cells cost about the same as silicon cells on a wattage basis but there are high hopes for reducing the costs by about an order of magnitude for thin film cells.

IV. NARROW BAND SELF FILTERING DETECTORS (NBSFD)

From Figures 2 and 3 it is observed that the response of a photovoltaic device is significant only when α lies between 10 and 1000. Thus if a material has an absorption coefficient versus wavelength characteristic that varies rapidly in the above range over a narrow spectral range, then the response characteristic versus wavelength will also have a narrow wavelength response. Figure 7 shows experimental data for a selection of semiconductor materials of the absorption coefficient versus wavelength. It is seen that there are several materials (GaAs, InAs, InP, etc.) that have the steep characteristic that would give the narrow band response. These materials are "direct" semiconductor materials in which electron-hole pair production takes place without the generation of phonons. Essentially what happens is that the short wavelength photons are absorbed in the thin layer adjacent to the surface generating hole-electron pairs which recombine before diffusing to the junction and the long wavelength photons pass through the device without being absorbed. Only the narrow range of photons having the intermediate values of α are absorbed in the body of the device and give rise to a response. Thus the top layer filters out the short wavelength photons and hence and name for the devices. These devices are more useful than using a broadband detector with a multiple layer interference filter since the latter is orientation wavelength sensitive while the NBSFD is orientation insensitive (due to the large index of refraction of the materials) and thus can be used as wide field of view devices.

There are four ways of varying the wavelength response; material composition tuning, temperature tuning, junction depth tuning, and reverse bias tuning.

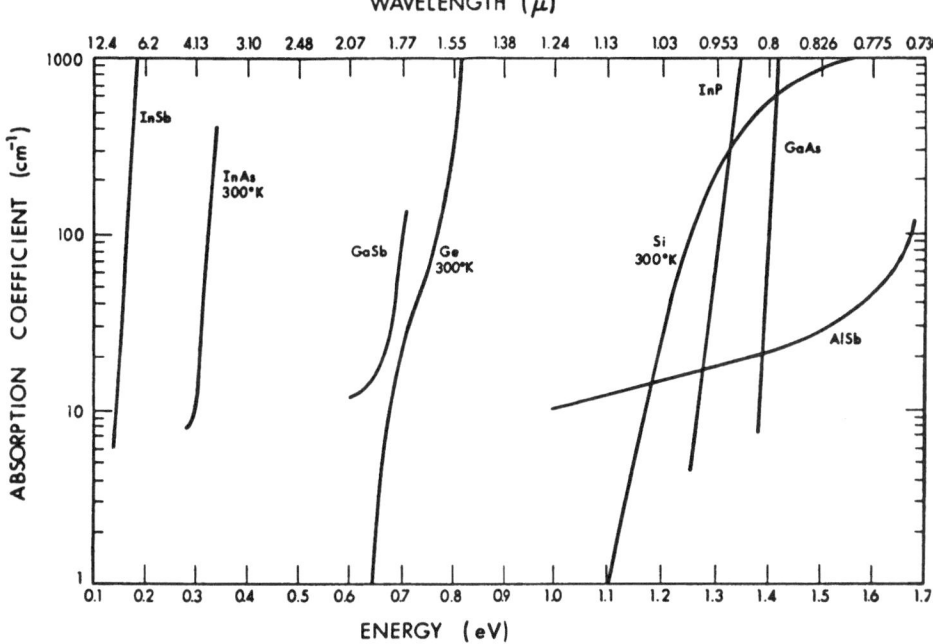

FIGURE 7. Absorption Coefficient versus Photon Energy for Various Semiconductor Materials

1. *Material Composition Tuning*

There are only a small number of binary compound direct semiconductor materials that can be used for NBSFD's. The use of only these materials would limit these devices to a few narrow wavelength bands. However by mixing these materials into terniary or quaternary compositions it is possible to vary the wavelength region of interest from 0.6 microns to 6 microns.

2. *Temperature Tuning*

By varying the operating temperature of the device one can vary the forbidden energy gap of the material and thus modify the wavelength region of response over a narrow range. For example using InAs, one could shift the response from 3.1 microns at 77°K to 3.7 microns at 25°C.

3. Junction Depth Tuning

As can be noted in Figures 2 and 3, a minor shift in response to smaller α and therefore longer λ can be made by placing the junction depth deeper in the material.

4. Reverse Bias Tuning

By designing the device with very high resistivity material for the top layer (region exposed to the direct radiation), it is possible to modify the effective junction depth by applying a reverse bias to the device. Upon application of the reverse bias, a space charge region is swept out through the top layer and two effects are noted. The first is that the total response will increase due to the wide space charge region and second, the effective junction is closer to the top surface and the peak response will occur at a slightly shorter wavelength.

All these techniques have been tested and found to work. Most of the work to date has been with GaAs since this direct semiconductor is readily available commercially in many impurity ranges. N-type GaAs is diffused with Zn to produce a p-n junction and the diffused p-region is the region unexposed to the direct radiation. These devices peak at 0.89 microns at 300°K and have a full width half maximum response of 150 Å. It is possible to shift the peak response by 50 Å by varying the depth of the junction below the top surface. In order to make detectors for narrow band response at 1.06 microns, crystals of $InAs_x P_{1-x}$ were grown where $x \cong 0.1$. Selection criteria for ternary materials were discussed in the lectures. The devices from this material peaked at 1.06 microns and have a full width half maximum response of 200 Å. Materials and devices have been made that operate at 2.7 microns and 4.2 microns, wavelengths which are of interest for the CO_2 absorption bands.

V. LATERAL PHOTOVOLTAIC EFFECT

Referring to Figure 1, a lateral photovoltaic response will be observed if a spot of light is permitted to fall on the device anywhere other than on the locus of equidistant points from the contacts (assuming a device with uniform spatial characteristics). The response can be shown to be proportional to the linear displacement from the above locus for distances small compared to the separation of the contacts. This voltage is developed due to the local production of electron-hole pairs which diffuse away from the region of their production giving rise to local electric fields. These fields give rise to equipotential surfaces which depend on the light intensity, boundary conditions, and region of illumination. When the contacts are symmetrically located and the light spot is not, then a potential difference will be found between the contacts. It can be shown that the illumination need not be a small spot and that the response is proportional to the integrated intensity and position of the centroid of the light spot. By having two pairs of contacts at right angles on a circular device, it is possible to have a vectorial output with the x and y components measured directly. These devices can be used with an optical system to measure the direction from which the illumination is coming. Also the devices can operate servo type electronics to track light sources since the output is reduced to zero when the light spot is centered on the device. Since the response is proportional to the intensity of the light, in many cases it is necessary to make an additional measurement of the transverse photovoltaic response (the light intensity) which can be used to normalize the lateral effect for determining the direction. Devices of this type are commercially available and are called radiation tracking transducers. They have been used in satellite applications primarily for tracking the sun.

VI. TRANSIENT EFFECTS

For most applications making use of the photovoltaic effect, the light falling on the detector usually is constant or changes very slowly compared to the time constants of the devices which are of the order of 10^{-5} to 10^{-9} seconds. However as more of these devices are finding their way into computer applications and laser detection with requirements for high speed of response, it is necessary to consider the transient response of photovoltaic devices. The time dependent equations of Section II can be solved for the transient response and yield transient solutions of the form

$$Q = \sum_{n=o}^{\infty} A_n e^{-\frac{[1 + \pi^2(2n + 1)D\tau]}{4W^2} \frac{t}{\tau}}$$

where the A_n depend on the boundary conditions and the other symbols were defined earlier. For solar cell type structures, it is easily shown that only the n = 0 mode contributes to the transient; the other modes decay extremely rapidly compared to the n=0 mode. In this case the transient has a time constant given by $T = 4W^2/\pi^2 D$ which is of the order of 10^{-8} to 10^{-9} seconds. For the NBSFD structure more than one term of the transient must be considered since W is much larger in this structure and the transient time constant is very close to τ in this case which again is of the order of 10^{-8} to 10^{-9} seconds. Deeper analysis of the transient response indicates that it is not possible to get faster response than the numbers above except by reducing the lifetime of the excess minority carriers. In doing this, the absolute response decreases as τ. Thus engineering trade-offs must be considered for the use of these detectors where response times faster than a nanosecond are required.

VII. MISCELLANEOUS DEVICES

In addition to the devices described above there are several others that make use of the photovoltaic effect or are special applications of those devices. One major class of devices are the p-n junction photodiodes. These devices are p-n junctions with a reverse bias applied across it and a load resistor in series with the diode. A current voltage plot showing the operation of such a device is given in Figure 8. In the dark, a small current flows and the voltage across the load is $V_R - V_D$ where V_R is the applied reverse bias and V_D is the voltage across the diode in the dark. When illuminated the current voltage characteristic is depressed (as indicated by the dashed curve) and the voltage across the diode falls to V_L with a resulting increase of voltage across the load to $V_R - V_L$. The voltage gain is thus $(V_R - V_L)/(V_R - V_D)$.

Small photovoltaic cells have been made in linear arrays with a common base as shown in Figure 9. Such arrays are being used in computer input and software equipment thereby eliminating mechanical fingers that erode computer cards and perforated tape. The latest semiconductor technology of planar diffusion and overlay insulated conductors have been applied to these arrays and also to two dimensional matrices with the latter being used as a solid state vidicon.

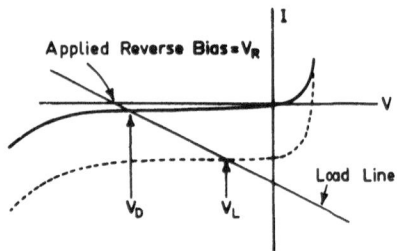

FIGURE 8. Photodiode Current-Voltage Characteristic

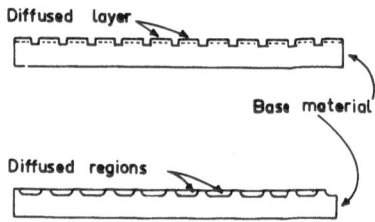

FIGURE 9. Photodiode Array Structures

For null detection of a small spot of light in one dimension it is possible to use two cells with similar characteristics and use the difference of the output to operate a servo control system. Likewise in two dimensions, the use of four rectangular cells will allow for control in both directions. So called split cells and quadrant cells make use of a common base and since the device is made at one time with the common material, the characteristics of the sub-devices are closely matched yielding devices with high accuracy.

Finally it should be mentioned that photovoltaic cells are usable as light meters and can be operated in two modes; one, in the short circuit mode where the current output is linearly proportional to the intensity and two, in the open circuit mode where the voltage output is logarithmically proportional to the light intensity.

REFERENCES

This paper has given a brief review of many subjects concerned with the photovoltaic effect. For the reader to find more details, a brief list of references is given of some useful papers in this field.

1. Solar Cells

M. B. PRINCE, J. Appl. Phys. 26, 534–540 (1955)
M. B. PRINCE and M. WOLF, J. Brit. Inst. Radio Engrs., 18, 583–595 (1958)
S. KAYE, Proc. Photovoltaic Specialists Conference — 1963, B11-1 — B11-15.
J. R. HIETANEN and F. A. SHIRLAND, Proc. Photovoltaic Specialists Conference — 1967, Vol. I, 179

2. NBSFD

J. W. BURNS et al., J. Appl. Phys., 38, 5388–5394 (1967)

3. Lateral Photovoltaic Effect

D. A. ALLEN, IRE Trans. Elect. Dev., 9, 411–416 (1962)

AUTHOR INDEX

SUBJECT INDEX

Lecture Notes in Computer Science 8125

Commenced Publication in 1973
Founding and Former Series Editors:
Gerhard Goos, Juris Hartmanis, and Jan van Leeuwen

Editorial Board

Advanced Research in Computing and Software Science
Subline of Lectures Notes in Computer Science

Subline Series Editors

Subline Advisory Board